Santa Monica Public Library
I SMP 00 1184233 T

American Operetta

American Operetta

From *H.M.S. Pinafore* to *Sweeney Todd*

Gerald Bordman

New York Oxford
OXFORD UNIVERSITY PRESS
1981

Library of Congress Cataloging in Publication Data

Bordman, Gerald Martin.
American operetta.

Includes index.
1. Operetta—United States. I. Title.
ML1900.B67 782.81'0973 80-20646
ISBN 0-19-502869-4

Printed in the United States of America

Preface

I'm well aware that operetta has become almost a dirty word. That saddens me. I'd rather sit through ten good operettas (unmiked and unamplified, of course) than through a musical comedy or revue. I suspect most Americans would, too, although most of them wouldn't want to call it operetta. So this book is dedicated to the proposition that until very recently we have continued to produce great new operettas, the more recent of which we have even started to revive, and that when we take the long-range view we will accept the all-too-long neglected operettas that once graced our stages and revive them as well.

In case you feel you've spotted an apparent inconsistency between the title of this book, *American Operetta*, and its subtitle, *From* H.M.S. Pinafore *to* Sweeney Todd, let me assure you the inconsistency is more apparent than real. *H.M.S. Pinafore,* along with *The Merry Widow* and several other importations that figure significantly in this book, may not be American works but the way Americans perceived and reacted to them and, most of all, the uses American writers made of them when they came to create their own theatre pieces will loom large in this study.

For their help I want to give special thanks to Hobart Berolzheimer, Lester Englander, Sylvan Greene, and Richard Traubner. The libraries of Millersville State Teachers' College and Lincoln University kindly made their excellent facilities available. The Theatre Collection at the Free Library of Philadelphia and the Library of the Performing Arts at Lincoln Center in New York were invaluable. I am grateful to Joellyn Ausanka for her careful typing of my sometimes illegible manuscript and to my copy editor, Kim Lewis, for ironing out the creases in my style. Most of all, my warmest thanks to my marvelous editor, Sheldon Meyer.

Yellow Wood Farm
Kirks' Mills, Pa.
December 1980

G. B.

Contents

American Operetta

1

Beginnings

Operetta.

How rarely we hear the word today. And what undertones of wistfulness or derision we often catch on hearing it. When was the last time an incoming Broadway musical boldly announced that it was a new operetta? Babies born the day of that announcement have long since reached maturity. Operetta as a description for a popular, contemporary musical entertainment has been banished into a theatrical limbo, and so, many would argue, has the very style of entertainment it connotes. Not so! Operetta is as alive today as it was a century ago, still being written, still being produced, and still being hailed by critics and playgoers. Only no one calls it operetta anymore. For the past forty years fashion has dictated that we call it a musical play—or something of the sort. What's in a name? A lot, obviously. Operetta by any other name is far more welcome.

Those who subscribe to the idea that a musical play is a modern and unique art form are equally assertive in consigning operetta to a decidedly lower rung of some esthetic ladder or a lower stage in the evolution of our musical theatre. They look upon it, when they deign to consider it at all, as a diversion for an age of innocence or for an era of elite culture well past or for a theatre still dependent on imported fare. They react with shock, and often with anger, to the suggestion that the modern musical play, from *Oklahoma!* to *Sweeney Todd*, is the direct and natural descendant of a distinguished line of operettas, starting as early as *La Grande Duchesse de Gérolstein* (*The Grand Duchess of Gérolstein*) and running through *H.M.S. Pinafore*, *The Merry Widow*, and *The Student Prince*.

There are substantive differences, of course, between these modern American operettas we call musical plays and the older ones. By the same token, however, there are important differences between, say, *Pinafore* and *The Merry Widow* or between *The Merry Widow* and *The Student Prince*. For that matter, there are marked differences between *Oklahoma!* and *Sweeney Todd*. Each generation of operetta has, after all, responded to its own day's theatrical requirements. Because of this, an element of "datedness" has set in the moment an operetta achieves its final shape. If we no longer write operettas as Gilbert and Sullivan or Franz Lehar or Sigmund Romberg did, neither do we write novels or construct buildings or ride from one place to another as we did in 1879 or 1907 or 1924. In time, even *Oklahoma!* and *Sweeney Todd* must be perceived as dated, to the extent that dated simply means belonging to an era.

To insist, however, that the differences between new schools and old necessarily demonstrate artistic advancement is historically, and artistically, shortsighted. For example, even the most ardent aficionados of older operettas will grant that Oscar Hammerstein and Alan Jay Lerner and Stephen Sondheim brought a refreshing poetry, elegance, and brilliance to their lyrics and, when they wrote it, to their dialogue. But we hear these writers' words with modern ears, which are attuned to elements of today's colloquial speech. Yet speech patterns change regularly, often swiftly. Fifty or one hundred years hence their very naturalness will sound in one respect or another unnatural—dated. Our earliest important lyricist and librettist demonstrates this all too poignantly. Yet, however stilted and artificial Harry B. Smith's lines sound today, they were much praised when they were new. Smith, in fact, was once so highly thought of that he was the first American lyricist honored with publication in hardcover of a compendium of his best work. So in one sense, judgment has to be perpetually reserved, or at least qualified. How hard it is to remember that today has a date, too.

Actually, calling an operetta a musical play is nothing new. Over the years operetta has gone by many names—opéra bouffe, comic opera, musical romance. It has often been labeled, specifically or generically, as musical comedy. Unfortunately, no clear-cut pattern of nomenclature has emerged. By design or by ignorance writers and producers have attached whatever description they saw fit to their shows, and so have created hopeless confusion from the start.

National origins are of little help. Opéra bouffe came to us through the French, and comic opera is English (from the French opéra comique). Less obvious is the German popularization of the term "operetta" itself. All three terms were employed haphazardly, sometimes for the same work. No doubt because it was so patently foreign, the term "opéra bouffe" quickly fell from favor in America, although certain puritanical considerations hastened its rejection. But the expression "comic opera" served every bit as well. It was an especially apt term for Gilbert and Sullivan's beguiling contrivances, and it hung on tenaciously for decades, long after the comic side of later comic operas had been subordinated to the romantic one. As a result, operetta—little opera—however serviceable as a blanket description, never attained universal acceptance.

One dictionary defines operetta as "a light, amusing, often farcical musical-dramatic work with an inconsequential plot, cheerful music and spoken dialogue." By way of distinction it views comic opera as "light opera usually having farcical dialogue between the musical numbers; loosely, musical burlesque," and opéra bouffe as "light comic opera with a preponderance of buffoonery or burlesque." Thus, it sees the differences as short, subtle gradations of declining artistic seriousness. For the record, the same dictionary defines *Singspiel*, the form out of which many scholars suggest operetta evolved, as "a semidramatic work, partly in dialogue and partly in song . . . it was often comic, had modern characters, and patterned its music on folk song with strictly subordinated accompaniment." If this last definition comes eerily close to describing latter-day operetta, none is truly comprehensive—to no small extent because operettas have been so varied.

Yet the definition of operetta itself is really not too far off the mark, provided allowances are made for its somewhat condescending nuances. Certainly operetta is light in comparison to opera, as well as light in its intention primarily to entertain. Indeed, "light opera" long served as a generic blanket covering all the variations of comic opera, operetta, and musical play. At its best operetta is often joyously amusing, although in the more romantic operettas comedy is frequently tangential. That much of this humor has been dreadful or totally lacking is a sad commentary on the skill of many librettists, not on any inherent flaw in the genre. It was a failing in many an older operetta and in modern ones as well. After *Paint Your Wagon*'s premiere critic Walter Kerr lamented, "Writing an *integrated* musical comedy—where people are believable and the songs are logically introduced—is no excuse for not being funny from time to time."

Unquestionably the most condescending injection in the dictionary definition was the adjective "inconsequential." It suggests the lexicographer subscribed to the idea that theatre must be as meaningful as it is entertaining. But pure entertainment has been an age-old function of the playhouse. The meaningful and the trivial always have existed side by side in the theatre. Until recently it has been a happy and welcome coexistence. So while many operetta plots have indeed been inconsequential, we should not inject pejorative connotations into the adjective. And, of course, operetta music has not always been cheery, except when the plot called for it. It also could be pensive, sad, coy, or exhilarating—whatever the moment demanded.

If, then, operetta cannot be precisely defined, it can be more readily distinguished from competitive forms. Since all operettas have plots, they are markedly different from revues. Furthermore, revue music has been much the same as that in musical comedy. Thirty years ago, in his pioneering study *Musical Comedy in America*, Cecil Smith observed, "Musical comedy may be distinguished from such other forms of entertainment as comic opera and burlesque [nineteenth-century burlesque] by its direct and essentially unstylized appropriation of vernacular song, dance and subject matter." Kings and queens rarely appeared in musical comedy, high society only intermittently. The proverbial man in the street, his girl, and their way of life were its province. Its vernacular songs were more an extension of recitative than they were lyrical elaborations, and ragtime and jazz were instantly at home in these works.

Even in attitude musical comedy differed from operetta, although often the difference was merely one of degree. Reviewing a 1918 production, Heywood Broun asked why the book of the show, "like most musical comedies, is full of cynicism about marriage while the lyrics . . . approach the self-same institution with reverence, truth and sweetness." Except for the earliest comic operas, operetta librettos generally took a far less jaundiced view of life than did those of musical comedy. Of course, many an operetta plot necessitated a certain underlying cynicism which occasionally surfaced in a biting remark or two. But as a rule, life, for all its uncertainties and perversity, was accepted at face value. The unpleasantness of the villain, if there was one, was carefully controlled, and gratuitous unpleasantness was all but unheard of. When random bits of cynicism or sarcasm were inserted, they were usually given to the comedian, often in a more or less extraneous comic song. It was clearly meant to serve as momentary comic relief. In a sense, then, the inconsistencies that Broun perceived in musical comedy were reversed in operetta.

Just when did comic opera or operetta or opéra bouffe move out of the opera house and onto the legitimate stage? An exact date is impossible to pinpoint and need not concern us here. Regrettably, our early theatrical records are so filled with gaps that we cannot be certain just when operetta as we would recognize it first played an American stage, or exactly what term was attached to it. By the 1790s, theatrical troupes up and down the American seaboard were advertising some offerings as musical farce, musical comedy, operetta, and comic opera. In 1793, an anonymous correspondent in the New York *Journal*—he or she used the name "Amateur"—wrote feelingly of "the good old Comic Opera of Love in a Village." Whatever the nature of the period's musical comedies and musical farces (and they were generally short afterpieces), its operettas and comic operas were assuredly what we now would call ballad operas, that is, plays for which substantial numbers of lyrics were written. At first these lyrics were set to already popular songs, although later ballad operas offered original music as well.

These English ballad operas and native imitations, then, were the first full-length lyric fare offered by American theatres. European operas soon followed. Spectacles, often labeled grandly as oratorios, also found wide audiences. In time, musical productions offering a mélange of styles, with elements borrowed from ballad operas, traditional operas, ballet, and spectacle, began to emerge. Advertisements displayed a wild assortment of descriptions for them—operatic extravaganza, musical comedietta, comic musical, and, in a few instances, musical comedy and comic opera. But the nomenclature was as haphazard as the musical assemblages themselves, and, understandably, no single name took hold.

Conceivably, the immensely popular English operas that appeared from the mid-1840s to the early 1860s smoothed the transition from the opera stage to more popular stages. They had a consistency of tone, easy melodies, and floridly romantic plots. Their readily accessible music gave the man in the street songs he could whistle. "I Dreamt That I Dwelt In Marble Halls" from William Michael Balfe's *The Bohemian Girl* and "Scenes That Are Brightest" from William Vincent Wallace's *Maritana* were among the "pop" songs of their era. Sung at every turn, they helped lure playgoers into the theatres; sung decade after decade they ensured revivals of both pieces all through the nineteenth century and even, on occasion, in the twentieth century.

But for early Victorian playgoers the stories these works told were equally alluring. Almost a hundred years later, in a revue skit, great comedienne Beatrice Lillie played an actress at the reading of a drama. When told the play will deal with a man and a woman, she dismisses it out of hand for having "too much plot." No such reservation crossed the minds of early audiences at *The Bohemian Girl, Maritana,* or the other musical triumphs of their era. Complications galore were interwoven into stories built largely on stock motifs and propelled by vigorous, unsubtle action. The stock motifs, many of which remained popular until a fashion for relative simplicity set in during the teens of the twentieth century, included lovers of unequal social rank, mistaken identity, and disguises. Many had served faithfully as standard story components for theatricals since Elizabethan times. Employed unsparingly, they gave plays an arch, unnatural tone and thus afforded the sort of wildly escapist fare we still associate with Victorian theatregoing.

The plots defy retelling in a sentence or two. For example: The hero of *The Bohemian Girl* is Thaddeus, a Polish nobleman and rebel, living in exile among a band of gypsies led by one Devilshoof. When Arline, daughter of Thaddeus's arch-enemy Count Arnheim, is attacked by a stag, Thaddeus rescues her and, accompanied by Devilshoof, returns her to the count. The count lets Thaddeus go but imprisons the gypsy. Devilshoof escapes, kidnaps Arline, and takes her to his encampment. Twelve years pass. Arline, still living with the gypsies, has fallen in love with Thaddeus, thereby inspiring the ire of the gypsy queen, who also loves him. The queen gives Arline a stolen medallion and then arranges for her to be captured by the count's men. Arline is tried for theft and convicted. However, the count recognizes the scar the stag had inflicted. Grateful at having recovered his long-lost daughter, he offers her anything she wants, but is appalled at her demand that she be allowed to marry Thaddeus. He rejects the idea of her marrying a lowly gypsy. Hurt by the count's arrogance, Thaddeus discloses his true identity. When Arnheim learns of the hero's noble origins, he relents and consents to the marriage. The furious gypsy queen attempts to shoot Arline at the wedding, but the bullet ricochets, killing the queen instead.

If, then, in conception and texture these English operas were somewhat grander than latter-day operettas, their uncomplicated lyricism and earnest romanticism pointed unwittingly toward generations of operettas to come.

By 1866 *The Black Crook*'s runaway success made American producers and theatre owners aware of the vast financial rewards to be reaped from musicals. The show set its tale of a man who sells his soul to the devil to music that ranged from theatrically inflated marches, galops, and waltzes to the most trite of barroom ditties. The entertainment was mounted with a baroque splendor heretofore unknown on Broadway and peopled with a cast of hundreds, including dozens of beautiful, buxom coryphees in pink tights. However, for all its opulence, *The Black Crook* was a crude, almost barbaric hodgepodge, little able to offer esthetic instruction to future composers and librettists.

But in the following year, 1867, Jacques Offenbach's *La Grande Duchesse de Gérolstein* swept New York off its feet. As one theatre historian noted, " 'Voici le sabre de mon père' and other airs of *La Grande Duchesse* were hummed, whistled, played until the ear was worn to shreds." Newspapers and tradesheets of the time confirm that he was hardly exaggerating. Opéra bouffe had arrived, and with it assuredly came operetta as we know it today. How well contemporaries understood it is moot, for these early French works were initially offered to America in French. As such, they could only appeal to a limited audience, however familiar their melodies. Nonetheless, they were joyously welcomed by the cognoscenti and by critics. The era's leading theatrical tradepaper, the *Dramatic Mirror*, insisted opéra bouffe was the highest order of popular musical theatre, arguing that French composers alone took care not to sacrifice the best elements of plot and characterization while creating sparkling, meritorious music.

If a language barrier prevented many Americans from fully relishing how "bouffe" these opéra bouffes were, no doubt the joyously bubbling music and captivatingly gay performances provided some encouraging clues. In France, of course, the comedy was totally understood and was responsible for no small share of Offenbach's early success. Actually, many of Offenbach's classics had become established favorites in Paris long before the premiere there of *La Grande Duchesse*. *Orphée aux Enfers*, *La Belle Hélène*, *La Vie parisienne*, and *Barbe-Bleue* had all preceded it. They brought with them to France's popular lyric stage a delicious comic spirit, reveling in the absurdities and excesses of the Second Empire's politics and high life. Inevitably, much of the fun had to be veiled. Politicians and hierarchies were as touchy then as now, and far more dangerously powerful. Thus ancient pantheons and time-cherished legends had to be employed as surrogates.

At heart, *La Grande Duchesse* was a spoof of militarism (so much so that French censors made the librettists change some all too specific jabs at famous French military campaigns). Americans, however, probably saw it as much as a devastatingly funny account of infatuation and its consequences. The love-obsessed grand duchess falls in love with a common private and instantly makes him a corporal. When she hears that he remains true to a peasant girl, she elevates him to lieutenant, hoping his new station will alter his affections. A general, eyeing the lieutenant's peasant girl for himself, attempts to humiliate the new officer, but this only prompts the duchess to make the lieutenant not merely a general but commander-in-chief. By some comic opera miracle the new general vanquishes the duchess's enemies. Her joy turns to horror when, as a reward, the general insists on marrying his beloved peasant. The duchess strips him of his rank and honors, bestowing them on a new infatuation, only to discover her latest dreamboat is married and the father of six children. He too is demoted. In the end she must marry the man her courtiers wanted her to marry in the first place.

The authors of the show may have been having some fun by giving Gerolstein, a tiny village in Germany, a grand duchess, and clearly they were out to ridicule the myriad and minuscule principalities that dotted the German countryside. The man whom the duchess finally weds is Prince of Steis-Stein-Steis-Laper-Bottmoll-Schorstenburg, a name larger than the imaginary principality itself and one that might pass for a contemporary German law firm.

Musically, the operetta left little to be desired. One writer has characterized Offenbach's score as "vital, sparkling, with fleeting moments of elegant tenderness," yet on the whole "always maintaining a necessary balance." Certainly none of the major singers could complain of being neglected. The couplets the composer set to music for the blustering general and the rondeau he gave to the befuddled private are among the highlights. However, the opéra bouffe's best moments inevitably fell to the duchess. With what was no doubt intentional irony in so anti-military a piece, much of her finest material was distinctly martial, notably "Ah! que j'aime les militaires" and the famous sword song. On the other hand, her great finale, "La Légende du verre," was appropriately more festive than warlike, while her earlier confession of love, "Dites lui qu'on l'a remarqué distingué" was caressingly gentle.

An unfamiliar language, indirection, and an ignorance of nuances satirized all presented obstacles to Americans. But fundamental humor and good will, coupled with such melodic, ebullient music, obviously got across enough of the intent and alerted Americans to the fact that operetta could be fun-loving and lighthearted just as successfully as it could be earnest. This undoubtedly eased the way for Gilbert and Sullivan a decade later.

The tide began to turn against opéra bouffe in 1871, when Offenbach's *The Princess of Trébizonde* was offered in English. Audiences and critics not versed in French realized for the first time how filled with double entendres and suggestive comments both the lyrics and dialogue were. Outraged, papers such as the New York *Herald* called for public condemnation. Nor were they content to assail the texts alone but, pronouncing guilt by association, now dismissed even the music as belonging to a "lower school."

Meanwhile, homegrown operetta had begun to appear on American stages. Probably the best known of the early composers was Julius Eichberg, a German who had trained in Brussels and Geneva, winning prizes both as a violinist and as a composer. He emigrated to New York in 1857 and moved to Boston two years later. All four of his operettas were given their first performances in Boston. Sadly, if they demonstrate his lack of growth and his failings as an artist, they also typify the shortcomings of early American operetta. His most famous work was unquestionably *The Doctor of Alcantara*, set to a text by Benjamin Woolf. The premiere of the show took place on April 7, 1862. During the 1866–67 season, the season before *La Grande Duchesse*'s American unveiling, three companies played it in New York.

Woolf's story was the sort that European opera had long used and which would become a commonplace in later American comic operas. It revolved around two young lovers who are ordered by their fathers to marry mates they have never set eyes on. Only after a series of highjinks do the lovers learn they themselves are the very mates their fathers have selected. Eichberg's music was as serviceable as

Woolf's plot—tinkly, germane, but derivative and in no way pleadingly melodic.

In short order, Eichberg followed *The Doctor of Alcantara* with *The Two Cadis*, *A Night in Rome*, and *The Rose of Tyrol*. Whether they had been composed earlier and were mounted to capitalize on *Alcantara*'s widespread acclaim cannot be readily ascertained.

Certainly *The Two Cadis*, first played at Chickering's Hall on March 5, 1868, was a far less sophisticated endeavor. It employed only four principals—a soprano, a tenor, a baritone, and a bass—and a small male chorus to unfold a simple tale of two roguish magistrates who abuse a vizier's son and are made to pay the price. A love interest between the young man and the niece of one of the magistrates provided some elementary complications. Elementary might also well describe its music, always allowing for Eichberg's technical skills. At one point Eichberg merely interpolated Beethoven's Turkish March for atmospheric background. Eichberg's own material was almost naïve compared with the brilliance of the best opéra bouffe scores or with the Sullivan scores that would appear ten years later. Even contemporary critics were aware of its limitations. One Boston critic noted, "The music, if not always original—least so in the set airs, most so where music and humor seem to spring up unconsciously and unpretendingly together—is very bright and pretty, and the concerted pieces, Trios and Quartet, very ingenious and effective."

"Where music and humor seem to spring up unconsciously and unpretendingly together"—what a tantalizing hint that integration of song and story was a consideration even in these early, prototypical offerings. If this anonymous critic was unknowingly touching on future aspirations, he and the original vocal score illustrated how quickly an infuriating confusion of terminology beset the genre. The review was headed "Mr. Eichberg's Operetta." The vocal score's title page called the piece a comic opera, apparently unconcerned that the plot synopsis described the stage picture "at the commencement of the operetta." Similarly, early editions, in the 1860s, of *The Doctor of Alcantara* labeled the work an opéra bouffe, while editions from the 1870s onward, no doubt responding to the disfavor into which opéra bouffe had fallen and acknowledging the acclaim accorded Gilbert and Sullivan, all termed the work a comic opera.

Eichberg's operettas and a handful of others served as transitional pieces. They were accepted by popular touring opera troupes and performed in repertory with the lighter operas these companies played from town to town. Caroline Richings's entourage typified these wandering minstrels. Miss Richings was the adopted daughter of a celebrated early-nineteenth-century actor, Peter Richings, and served her apprenticeship in his dramatic offerings. She rocketed to fame on April 14, 1862, when she sang the lead in her father's mounting of Balfe's *The Enchantress*. For the remainder of her career she was identified with the role of Stella, and *The Enchantress* was rarely absent from her repertory. She added *The Doctor of Alcantara* to her troupe's schedule in 1867, bringing it to New York with *The Bohemian Girl*, *Maritana*, and operas of a like order.

Not all American operettas were even as artful as Eichberg's relatively simple compositions. Henry Clay Barnabee, who became one of the stellar performers in late-nineteenth-century American operettas, recalled an example of these rudimentary pieces in his charming autobiography. Barnabee had sung the basso role in the Boston premiere of *The Two Cadis*. By the end of the same month he was in Portsmouth, New Hampshire, for two performances of *The Haymakers*. By an author whom Barnabee remembered solely as "a farmer," the work "called for two farmer's daughters, soprano and contralto, and two hired men, tenor and bass, to fall in love with them, and incidentally to engage in a series of musical mix-ups, such as duos, trios and quartettes." Financial considerations prevented Barnabee from remaining with the production (which, as he recalled, cost about a hundred dollars) when it left Portsmouth to tour other New England towns. Wherever it went, it was advertised as an "operatic cantata," apparently in reaction to the same anti-theatrical prejudices and pruderies that soon assailed opéra bouffe.

Fears of encountering just this sort of prudish opposition may have prompted the producers of another operetta to use a similar designation when they brought it to New York seven years later. The first American production of *Trial by Jury* arrived in New York on November 15, 1875, billed as an "original and dramatic cantata." The billing deceived no one. In any case, the piece received a lukewarm welcome and was hastily withdrawn.

But matters changed once and for all in 1878 when a second Gilbert and Sullivan work arrived. *H.M.S. Pinafore*, English though it was, marks the real beginnings both of Americans' love for operetta and of American musical theatre itself.

2

H.M.S. Pinafore: The Rise and Early Evolution of Comic Opera

It seems such innocent fun, *H.M.S. Pinafore* does—clever, delightful, obviously artful. Its hilariously cockeyed logic and its joyful melodicism are instantly embraceable and remain lovable after repeated hearings. Familiarity breeds admiration. Yet to call it world-shaking might appear unthinkingly hyperbolic. Nevertheless, in the world of English-speaking theatre it stands, arguably, as the most important musical ever written. *H.M.S. Pinafore*, along with two light opera companies it helped establish, determined the course and shape of the popular lyric stage in England and America for the final quarter of the nineteenth century. More cogently, this English masterpiece marked the real beginning of the American musical theatre as we know it to this day. Without *Pinafore*, conceivably, there might have been no Victor Herbert, no George M. Cohan, no Jerome Kern, no George Gershwin, no Richard Rodgers.

Neither Gilbert nor Sullivan was to the theatre born. William Schwenck Gilbert (1836–1911) was a descendant of famous soldier and navigator Sir Humphrey Gilbert and the son of a part-time novelist. Unable to realize his early ambitions for a career in law or the military, young Gilbert turned to writing comic verse over the signature "Bab." He also became drama critic for the *Illustrated Times*. His first play was a spoof of Donizetti's opera *L'Elisire d'amore* and an immediate success. He continued to write plays, serious and comic, until 1891. The most successful was, assuredly, *Engaged*, first played in 1877 and today still revived at intervals. Gilbert's association with Sullivan began in 1871 with a quick failure, *Thespis*; or, *The Gods Grown Old*. Four years later their one-act *Trial by Jury* received a far more encouraging reception, at least in England. In 1877 *The Sorcerer* premiered. But it was *H.M.S. Pinafore*, mounted at London's Opera Comique on May 25, 1878, that gave them the success they both so anticipated. A "master of metre," Gilbert deftly juggled quips, paradoxes, and absurd dilemmas within sophisticated and strict forms. One early biographer went further, noting, "Even for the music of the operas he deserves some credit, for the rhythms were frequently his own (as in 'I have a Song to Sing, O') and the metres were in many cases invented by himself."

Arthur Sullivan (1842–1900) came from a musical family. His father was a bandmaster, and young Sullivan is reputed to have been able to play all the wind instruments in the band by the age of eight. (This connection between military bands and composers of operetta will continue to crop up. Curious and largely unexplored, it deserves further study.) Endowed with a fine boyish voice, the future composer spent time as a chorister of the Chapel Royal in London. He then studied music at the Royal Academy in London and in Leipzig. The influence of German romanticism was thereafter to remain an important part of his music. His early compositions included a cantata, a ballet, art songs, and background music for a production of *The Tempest*. In 1867 he first ventured into comic opera, writing *Cox and Box* and *The Contrabandista* in collaboration with F. C. Burnand. Despite the success that later accrued to him and Gilbert, Sullivan's participation was somewhat grudging. All through his life he aimed at being accepted as a serious composer, a station he felt denied him because of his writing operetta. Yet his loftier works, whether successful or not in their own day, are forgotten, and apart from two notable exceptions—

"The Lost Chord" and "Onward, Christian Soldiers!"—it is his work with Gilbert that survives.

The drolleries of Gilbert's libretto and lyrics for *Pinafore* were typical of what he had offered his associate and his audiences both in the earlier Gilbert and Sullivan gems and in the later ones. The plot was deceptively simple—startlingly so for its era. A common sailor, Ralph Rackstraw, loves Josephine, the daughter of the *Pinafore*'s Captain Corcoran. Only the disparity in their social positions keeps them apart. Their problem is neatly solved when an old bumboat woman, Little Buttercup, discloses that she had once been both Ralph's and the captain's nursemaid—and had mixed up the two. Ralph is really Corcoran and Corcoran, Ralph. The complications in the plot were little more than having Sir Joseph Porter, the First Lord of the Admiralty, fall in love with Josephine, and the villainous Dick Deadeye squeal on Ralph and Josephine's planned elopement. The real complexities often derived from the characters' attitudes rather than from real turns of plot. Thus, Josephine bewails, "I hate myself when I think of the depth to which I have stooped in permitting myself to think tenderly of one so ignobly born, but I love him!" Believing that Josephine is afraid to acknowledge her affection for him because of his lordly position and her relatively low social standing, Sir Joseph argues that "love can level ranks." The assurance proves doubly ironic, for not only does it cause Josephine to view Ralph more positively but, after Josephine is shown to be a mere seaman's daughter, it reveals Porter as a snobbish hypocrite. He wheedles his way out by suggesting that love does level ranks "to a considerable extent, but it does not level them as much as that." In essence, then, Gilbert was having fun with a basic motif of operetta, the love between low-born and high-born. The length to which Gilbert could take absurdity, and get away with it, was implicit in Buttercup's confession, for Ralph and Corcoran could never have been babies at the same time.

Such absurdity underscored the fundamental artificiality of much of Gilbert and Sullivan's work, and, for that matter, of truly comic opera. In later, more romantic operettas, this note of conscious artifice all but disappeared. The later operettas attempted to present stories which, in varying degrees, audiences could accept as little more than heightened or stylized reality, peopled by characters with whom audiences could, again in varying degrees, empathize. Significantly, in these later works the most apparent bit of artificiality was often the comic relief, or, more specifically, the comedian—early on, frequently a grotesque comedian, afterwards a dialect comedian. But in comic opera a cultivated, neatly refined artificiality was maintained throughout, so that playgoers were always aware they were watching a stage picture.

Probably no better example of this artificiality exists than the patter songs that Gilbert and Sullivan perfected. The speed with which these tongue twisters had to be articulated spotlighted their unreality. And their sense was, of course, the most wonderful nonsense. However startling the career of William H. Smith, who rose from bookseller to the head of Queen Victoria's navy, and after whom Gilbert patterned Sir Joseph, it takes on an unworldly, lunatic logic when Sir Joseph admonishes his listeners,

> Stick close to your desks, and never go to sea,
> And you may all be Rulers of the Queen's Navee!

The fatuously proud Major-General Stanley of *The Pirates of Penzance* goes Sir Joseph one better, admitting,

> For my military knowledge, though I'm plucky and adventury,
> Has only been brought down to the beginning of the century;
> But still in matters vegetable, animal and mineral
> I am the very model of a modern Major-General.

Hilariously sane and demented at one and the same time, these musical communiqués assured audiences they had been transported to neverland.

But even the most brilliantly sustained comic opera would not want to be unwaveringly comic. Some element of believability was required. For these moments Gilbert often created remarkable, straightforward lyrics (with, admittedly, an occasional comic intrusion) that allowed Sullivan to set them to traditional forms. The sea chanty that opens *Pinafore* is a rousing example. *Pinafore* also offers a barcarole and a glee sung a capella. *The Mikado, Ruddigore,* and *The Yeoman of the Guard* include madrigals in their musical programs.

Patience offers the hymnlike "I hear the soft note of the echoing voice."

An unerring balance of wit and sentiment, all set exquisitely to music, then, was what *Pinafore* offered American playgoers in 1879. Statistics and history underscore any claim of *Pinafore*'s importance. In 1879 New York's theatrical dominance was still years in the future. Bustling, enlightened Boston presented the American premiere of *Pinafore* on November 25, 1878. San Francisco followed a month later, while Philadelphia offered its version the first week of 1879. In all three cities rival companies were soon active. New York had to wait until January 15 to chortle over Gilbert's wit and hum Sullivan's irresistible tunes. But having heard the work, the city wouldn't let it go. At one point in the season three houses were playing different interpretations. Since New York had only twelve major legitimate theatres in 1879, a modern equivalent would be seven or eight companies all performing the same hit at once. By the end of the 1878–79 season every major New York theatre had at some time during the year played host to a *Pinafore* company. Before *Pinafore* appeared, fewer than a dozen musicals were offered on Broadway each season (counting an occasional opéra bouffe repertory company as a single offering). *Pinafore* changed that dramatically. Starting in the very next theatre season—and for a half-century thereafter—Broadway never had fewer than twenty or thirty musicals a year.

Despite all the excitement and huzzahs, a number of perceptive, sensitive playgoers were not overwhelmed. They appreciated Gilbert's literate, biting wit and Sullivan's thoughtful, memorable score. Moreover, they were aware how carefully *Pinafore* integrated words and music. Songs developed the plot and moved that development along. Such integration was a rare virtue on popular lyric stages at the time. No, what offended many theatregoers were the shoddy presentations *Pinafore* suffered. With Sullivan's brilliant orchestrations not yet available, most pit bands blared forth with horrendously jangling, home-contrived instrumentations. On stage, matters were no better. There was little sense of the stylization necessary to put the piece across properly. Worse, the text often was not adhered to faithfully. Local jokes frequently were inserted, and, just as frequently, popular airs were added to spice up the score. Although no one could foresee it, these interpolations were an ominous foreshadowing of the treatment many imported operettas would receive in ensuing decades.

One objector to such practices was a Boston newspaper, which suggested that the city had the wherewithal to present an "ideal" version of the comic opera. Authorities at the Boston Theatre read the article, approved its sentiments, and approached a local lady, Miss Effie H. Ober, who served as agent for several respected artists. The theatre's management asked her to form a company that might indeed give Boston an "ideal" interpretation. Miss Ober agreed and set about her assignment with enthusiasm and taste. She enlisted Myron Whitney, a celebrated oratorio singer, to play Captain Corcoran, Tom Karl for the role of Ralph Rackstraw, Mary Beebe as Josephine, George Frothingham as Dick Deadeye, Isabella McCulloch to portray Buttercup, and, for Hebe, the beautiful actress whose career was to be so tragically foreshortened, Georgia Cayvan. Best of all, to play the part of Sir Joseph Porter, Miss Ober signed on the great comic bass-baritone who would be the mainstay of the troupe for most of its quarter-century history, Henry Clay Barnabee. Barnabee, born in 1833 in Portsmouth, New Hampshire, had little formal musical training and to the end of his days insisted he could not read music. Yet from the time he left his job as a clerk in a dry goods store to entertain in local lyceums he revealed an impeccable gift for comedy, a fine voice, and unyielding artistic integrity.

Barnabee must have been instantly delighted with his new associates. Nothing was spared to give the work a sumptuous yet tasteful mounting, alhough whether Sullivan's orchestrations were employed is unknown. As the Boston Ideal Opera Company, the troupe gave its first performance on April 14, 1879. Audiences and critics alike immediately recognized the players for what they were—the finest artistic ensemble ever presented on America's popular lyric stage. The company was rewarded with a nine-week run, a Boston record at the time.

The quality of the performance led an unidentified critic for the Boston *Journal* to see exalted aspects in the composition of *Pinafore* that had heretofore eluded him. How well taken his arguments were is moot; the points he made demonstrated a relationship that was not totally lost on his contemporaries and that indirectly had a marked significance on the immediate future of our musical stage:

> Many people will make the mistake of considering "Pinafore" a burlesque, and that word is, unhappily, suggestive of rather objectionable surroundings. Several good companies have badly mangled "Pinafore" by treating it as of this class of composition, for while irresistibly comical, it is not *bouffe*, and requires to be handled with great care lest its delicate proportions be marred and its subtle quality of humor lost. Moreover, it is constructed on principles foreign to opera *bouffe*. Its method is that of the classical operas, and its most exquisite satire lies in its imitation of the absurdities of such compositions—and that they have absurdities nobody who compares their artificiality with the occurrences and emotions of real life will be disposed to deny. Solos and choruses, duets and quartettes, and the other arrangement of voices which occur in classical operas are here duly presented, and one follows the other according to rule, and the whole motive of the work is directed by long-established precedent.

Precisely what prompted the critic to make this connection with grand opera must forever be lost, but conjecture helps explain the shape of this era's operettas. For one, the critic may have shared in the puritanical revulsion against opéra bouffe. By discrediting its influence on Gilbert and Sullivan, he was aligning himself with a number of anti-theatrical clerics who, in a notable about-face, had hailed *Pinafore* as good, clean family entertainment. Thus he was casting one more powerful vote for the tenor of comic opera to come. Another possibility is that he was moved by the performers' credentials. Several had appeared in Boston and elsewhere with major opera troupes. By thus obliquely underscoring the nature of the casting, he was unwittingly reinforcing the nature of future comic opera casting.

Or the critic may simply have been responding to the company's name. How telling that these New Englanders, who were knowingly mounting a comic opera, chose not to call their troupe a comic opera or light opera company. At this early date they could not have foreseen how long-lived their entourage would be or how broad-ranging its repertoire. For all their excellence, calling themselves an opera company at this point was a little pretentious, even if it was also a little portentous. Of course, the critic may merely have been recording an observation he made as he sat through the premiere, although one can only wonder what rule book he had in mind.

Excursions into grander opera were months away for the Boston Ideals, and when they decided to capitalize on their initial success by mounting a second work, their choice was almost as significant as *Pinafore*, albeit again no one at the time could understand that significance. Franz von Suppé's *Fatinitza* had first been presented to New York on April 14, 1879, the very night the Ideals were unveiling their *Pinafore*. Nowhere near the raging success *Pinafore* had been, the work nevertheless initiated a vogue for Viennese operetta at practically the very moment Gilbert and Sullivan's opus was inaugurating the reign of English comic opera. These two traditions—English comic opera and Viennese operetta—would rule America's musical theatre for the rest of the century, and American imitations would permit native artists to take their first, tentative steps along beaten paths. While historical and ethnic affinities may have determined that America's earliest efforts would cling more closely to English than to Austro-Hungarian ways, both traditions were influential. Shortly, English, Viennese, and American works would compete along that length of Broadway then known affectionately as "the Rialto."

One reason that *Fatinitza* failed to match *Pinafore*'s appeal may have been the boudoir-farce elements in its story. In this respect it prefigured the stories of future Viennese operettas, which were so often framed on accounts of somewhat illicit romantic dalliances. Just why the same prudes who railed so vociferously against opéra bouffe held their peace with regard to these German efforts is hard to say, although quite possibly they detected a certain leer or salacious wink in the French that they missed in the German.

Several subplots enlivened *Fatinitza*, but its principal story began during a lull in the Russo-Turkish War, with Lieutenant Vladimir Michailoff mooning over the absence of his beloved Lydia, ward of the boorish General Kantschakoff. Vladimir confesses to his comrades that once, desperate to be near Lydia, he dressed himself as a woman, took the name of Fatinitza, and applied for a post as Lydia's companion. His stratagem came to naught when the general fell madly in love with the disguised soldier, so he was forced to flee. To pass time, the officers decided to stage a play based on the incident. Vladimir agrees to impersonate himself. He is no sooner in his costume than Kantschakoff appears and once again begins his pursuit. Lydia's arrival at the encampment further complicates matters as does a raid by the Turks. The Turks seize Lydia and Fantinitza, taking them to the governor's harem. Kantschakoff hurries off to rescue his ward and his inamorata, but Lydia and Vladimir escape without the general's assistance. Back home, the general demands Lydia wed an aging prince. He will not hear of his ward marrying the lieutenant. Vladimir offers a compromise. If he can produce Fatinitza, Lydia will marry him; if not, she will marry the prince. Lusting as he does for Fatinitza, Kantschakoff agrees. When Fatinitza appears, the general lunges for her, but Vladimir pulls off his wig and the general realizes he has been gulled. He accepts defeat in time for a happy curtain.

If a suggestion of bawdiness may have deprived *Fatinitza* of its full measure of success, two other shortcomings no doubt hurt it as well. In its original and in translation it could not match Gilbert's brilliant libretto and lyrics nor Sullivan's irresistible music. Even in Germany today von Suppé's operetta scores are rarely heard. It would remain for Johann Strauss to do for Viennese operetta what Offenbach had done for French opéra bouffe, and Sullivan for English comic opera.

By no small coincidence, this same 1878–79 season that witnessed the American premieres of *Pinafore* and *Fatinitza* also witnessed the premieres of the first great Harrigan and Hart vehicle and the first of the successful "farce-comedies," those curiously slapdash, song-filled vaudevilles. Ned Harrigan and Tony Hart brought to the stage unflinching, if loving and humorous, glimpses of New York low life, centering naturally on their own Irish brethren, but treating poor blacks, Italians, and Jews as well. For the most part their plays were well plotted, with large casts, and filled with homey David Braham melodies for which Harrigan provided the lyrics. By contrast, the farce-comedies often had no more than six or eight players and virtually no plot. *The Brook* simply had its five entertainers go on an imaginary picnic, during which they divert themselves and their audiences with a collection of specialty routines and with songs culled from elsewhere. In retrospect, Harrigan and Hart and farce-comedy can be seen as unmistakable progenitors of the genre that would ultimately give operetta its fiercest competition, musical comedy.

Up to this point, at least in the United States, probably only Boston had seen operetta done with true panache. Affairs changed permanently in the next season—1879–80. By that time, of course, the success of *Pinafore* in London had secured fame and financial stability for Richard D'Oyly Carte's opera company there—a company still performing more than a hundred years later. Gilbert, Sullivan, and D'Oyly Carte's reasons for hazarding a risky, ofttimes unpleasant sea voyage to America were not governed by artistic altruism. Although these men were not above seeing their coffers filled with American dollars, they undoubtedly would have preferred to watch the money roll in while they were comfortably situated by their English firesides. But, in fact, little or no money was rolling in. Technicalities in the copyright law prevented their sharing in *Pinafore*'s American success. As a result, they determined to secure a copyright for their next work, *The Pirates of Penzance*, by offering its world premiere in New York. However, first they would show Americans the way to do *Pinafore*. Tactfully, they hired a few Americans for some of the principal parts. But sets, costumes, orchestrations, and many of the best artists were the same London had applauded. The original staging was carefully rehearsed, and no interpolated jokes or songs allowed. New York had its eyes opened.

The Pirates of Penzance failed to entice Manhattan's playgoers as wildly as *Pinafore* had. It ran a disappointing two months. Perhaps the public, lacking time to fully savor its excellences, could only see it as little more than a rewriting of *Pinafore*—another story deriving from a mix-up created by a blundering nursemaid. On the other hand, critics were overjoyed, not merely by the delights of the text, which one critic celebrated as "brighter, prettier and more artistic" than *Pinafore*, but for the imaginatively designed and carefully built sets, the tasteful costumes, the superior singing and superb ensemble playing. Here were performances that displayed not merely professional competence and occasionally personal genius, but offered a consistent, thoughtful stylization and artfully conceived stage pictures.

Three months to the day after the Englishmen initially performed for New York, the Boston Ideal Opera Company came into Niblo's as part of what was to be the first of its many national tours. Even at this early date New York critics had developed the haughty disdain they have long held for other cities' plays and players. It was clearly permissible to hail London's best efforts. London was a sort of sister city, the theatrical center of a great, related nation. But Boston? That was stretching courtesy. Yet these New England artists opened New York's eyes every bit as much as London's artists had. The *New York Times* was forced to confess that this latest *Pinafore* was "the most perfect, musically at least, that has yet been given in the City." Almost defensively, the *Times* and other papers in "the City" found fault with the physical production and the acting. But their complaints had a sniveling air to them, and quality would out. After a short while even New York's most chauvinistic critics made an exception for this troupe, allowing that it was a special joy. No doubt, too, the Ideals responded to legitimate criticism, improving their art as they moved onward.

Before long, praise for the Boston Ideal Opera Company or, as they were later called after some internal reorganization, the Bostonians, rang out across North America. The San Francisco *Examiner* wrote, "The company is too well known to need a recapitulation of its merits. Good looks, good voices, good action and good management make its work remarkably satisfactory." The Montreal *Herald* exclaimed, "The Bostonians have been together since the beginning of American lyric opera. It is not strange that they should be its best interpreters." Notices for the players reflected their excellence both as individual artists and as ensemble performers. A Kansas City critic said of Barnabee, "At all times he is a gentleman, and nothing in his quiet and Jeffersonian wit is ever vulgar or out of place." Often carping Lewis Strang, an early theatrical historian, could say of Jessie Bartlett Davis, "She has never been forgetful of the art of interpretation and of expression, and by means of her beautiful voice she has kept herself well in the lead among light opera contraltos."

In their quarter-century of existence the Bostonians built an impressive repertory that not only included most of the major works by Gilbert and Sullivan, von Suppé, and Offenbach, but extended to the lighter masterpieces of grand opera—Lortzing's *Czar and Carpenter*,

Flowtow's *Martha*, Donizetti's *Elixir of Love*, and Mozart's *Marriage of Figaro*, to cite a few. Besides Barnabee and Miss Davis, their ranks included Tom Karl and George Frothingham from the company's earliest days, William T. Carleton, Eugene Cowles, and Alice Nielson—all beloved and respected in their time. To the company's credit, they always toured with their entire repertory. When at their zenith they bravely offered untried works, these, too, were offered as quickly in one-night stands as in longer stays.

Inevitably, both the Bostonians and the D'Oyly Carte company raised standards of comic opera performance, although the Englishmen's influence was less direct. The D'Oyly Carte was the first to force New York critics to elevate their sights. They presented Manhattan reviewers a glorious example by which to judge all forthcoming musical offerings, and the Rialto's critics readily adopted a more demanding stance. As a result, when shows thereafter left New York to tour they often bore with them evidences of improving standards. The Bostonians' influence was more direct. As year after year they ventured farther afield to demonstrate their art, standards were quickly lifted and imitators were spawned. Some were very good, some dreadful.

These copycat companies have passed into theatrical history, and probably no one remains alive today who saw them. Yet their very names created an air of excited expectancy for a generation, providing widespread audiences with delightful escapist evenings and lifelong memories. A few troupes crisscrossed the nation; most remained regional. The Alice Oates Opera Company, the Solomon Opera Company, the Comley-Barton Opera Company, the Emily Melville Comic Opera Company moved wherever far-flung railroad networks would take them. On the other hand, the Pike Opera Company never ranged far from San Francisco, and the Chicago Church Choir Opera Company confined its art to the Midwest. New England supported an especially rich crop, besides the Bostonians. Boston's Castle Square Light Opera Company, under the aegis of young Henry Savage, soon became so successful that two branch companies were formed, one of which played two full seasons in Philadelphia while the other toured the East. The First Corps of Cadets entered the Boston arena a few years later, although it quickly moved away from comic opera into the new field of musical comedy. In 1900 Strang reminisced about yet another troupe:

> The Bennett and Moulton Opera Company was a frequent visitor to the small cities and large towns of New England. It played week stands with daily matinees, and it was, more than likely, the pioneer of "ten, twenty, thirty." I have every feeling of gratitude toward the Bennett and Moulton Opera Company, for it introduced me, at the modest rate of ten cents per introduction, which small sum purchased the right to sit aloft in the gallery, to all the famous old-time operettas.

And these bands of vagabond performers were but a few of the numerous old troupes singing and clowning on another century's stages.

Not all players were forced to become such wanderers. Alongside traveling repertory troupes, light opera stock companies developed. Local organizations, they characteristically presented each opera or operetta for a week at a time. In New York, John McCaull and Rudolph Aronson founded companies which maintained the highest standards and were often rewarded with extended runs. The noted comedian Francis Wilson described his sometime employer as follows:

> Colonel John A. McCaull was a soldier who fought all through "the late unpleasantness" on the Southern side. He was twice wounded, made prisoner and carried to Fort Warren, from which he was pardoned by Abraham Lincoln. Of Scotch-Irish extraction, he had all the impetuosity and pugnacity of his progenitors. He had been trained to the law, and defending some theatrical suit brought to him by Emily Melville, I believe, became interested in that branch of the theatrical profession with which he afterwards became allied. A proud man, he was swift to take offense: he could be a firm friend and a bitter enemy. His impulsiveness often warped his judgment.

Some colleagues have suggested that McCaull's initial contact with the stage was as attorney for John Ford of Ford's Theatre in Baltimore, but this aside, Wilson's portrait appears accurate. Wilson makes no mention of any musical training or pervasive musical background. McCaull entered the theatre, possibly drawn by its glamour, simply as a promoter. He called his troupe the McCaull Opera Comique Company, established several branches, and somehow earned the epithet "Father of American Comic Opera."

Aronson was also a promoter, but he brought to his efforts an extensive background in and fervent love of music. His dreams were Olympian when they were not Utopian. Late in life, after his careless management and misjudgment had cost him his career, he was still envisioning a vast "American Palace of Art," a sort of Lincoln or Kennedy Center decades ahead of its time. A New Yorker of German-Jewish extraction, he traveled extensively in Europe and fraternized with many of the leading composers and singers of his day. His first New York enterprise was a small concert hall. In 1880 he began to plan a second auditorium, one that could be used for both concerts and light opera, and one that would contain its own restaurant and shops. Moreover, borrowing from an idea he had seen in Europe, he conceived the notion of a roof garden where musicals could be moved when summer heat made an enclosed playhouse stifling. He named his pseudo-Moorish house the Casino (hoping, as well, to have some gaming rooms attached).

Although Aronson received substantial backing, his ambitious schemes had to be modified. And his slipshod planning was laid bare when the opening had to be postponed, and when even the delayed first night revealed the house was still so uncomfortably far from finished that the initial attraction had to be taken off and hustled to Philadelphia while remaining construction was attended to. Nevertheless, when the Casino was ready for regular patronage in 1882—just three years after the advent of *Pinafore* and the birth of the Bostonians—it marked the first time a theatre had been opened in America exclusively to present comic opera. The company Aronson assembled soon became almost as much an exemplar as the Bostonians.

It was Aronson who gave Johann Strauss most of his important early American mountings, including *The Queen's Lace Handkerchief*, *Prince Methusalem*, and *Die Fledermaus*. Yet curiously, Strauss's masterpieces found little acceptance in America during his lifetime. *Die Fledermaus* and *The Gypsy Baron* would not be truly popular until the twentieth century. Rather it was what we today would consider minor Strauss, operettas such as *The Queen's Lace Handkerchief* (*Das Spitzentuch der Königin*), *Prince Methusalem*, and *The Merry War* (*Der lustige Krieg*), that first won favor. These were not negligible compositions. The first gave us "Rosen aus dem Süden" ("Roses from the South"); the last, "Nur für Natur." Both songs were waltzes, and therein lay Strauss's genius—and the importance of his work to operetta, for it was Strauss who effectively brought personal dance music to the lyric stage. Of course, professionals might have performed Offenbach's can-can. His and von Suppé's galops and schottisches might have had their brief moment on the dance floors, and even a few of Sullivan's melodies could have been waltzed to. But Strauss's experience with the great dance orchestras in Vienna's glittering ballrooms gave him feeling and insight into the requirements of his era's social dancers. It must be remembered that this was not an epoch given to soft, caressing dancing. Dancing was more formal, more impersonal, more bravura, and Strauss's waltzes, as well as the rest of his operetta music, reflected this. It was vaultingly theatrical.

Aronson also brought New Yorkers Carl Millöcker's *The Beggar Student*, and, most successfully, Edward Jakobowski's now virtually forgotten *Erminie*. For shorter or longer stretches Aronson's company included Lillian Russell, Lilly Post, Mathilde Cottrelly (a superb comedienne and behind-the-scenes business manager), Berthe Ricci, Pauline Hall, and the great clown, Francis Wilson.

What were these companies like? Barnabee recalled that the Bostonians "had very much the aspect of a large family party out on a picnic jaunt." Clearly even the less elite, less long-lived troupes shared something of this sentiment. Large or small they annually issued company portraits, duly published by trade journals and clipped by admirers. The pictures, as much as the few surviving records, confirm the stability of the better bands. True, each year a few new faces smiled out at readers, but for the most part a hardy crew of loyalists were a company's mainstay. Barnabee's own twenty-year stint with the Bostonians was something of an exception, although George Frothingham and Jessie Bartlett Davis remained with the company almost as long. But then the Bostonians outlasted every other American troupe, bestriding the entire era of these repertory ensembles. Yet other groups elicited their own loyalties. In a notoriously unstable trade, performers often looked with both gratitude and pride at their seasons with particular troupes. The beloved comedian Jefferson De

Angelis noted he had remained with McCaull's Philadelphia company for three years and then moved on to a far longer stay at the Casino. Marion Manola boasted of remaining four years in McCaull's New York entourage, while Lilly Post could point to a six-year sojourn with the same band. These performers were leading figures in their great troupes, experienced, established actors and actresses. At the same time, lesser repertory and stock ensembles offered a fertile training ground for a succeeding generation of favorites. Out of the Castle Square Theatre troupe, for example, came Raymond Hitchcock, Frank Moulan, Lizzie MacNichol, and several other turn-of-the-century celebrities. Hitchcock also spent time with Bennett and Moulton, as did Della Fox.

Ironically, these repertory and, especially, these stock companies came to life just as social and technological changes were signaling the death knell of the dramatic stock companies that had flourished in the American theatre. In New York, for example, Wallack's, Daly's, and Palmer's had represented the pinnacle of American theatrical art, presenting the best new and old plays in the most beautifully meshed acting ensembles. But a growing population and the post-Civil War expansion of train travel made it economically more feasible to send a single show, specially cast, on a long tour after its New York run. "Original New York Cast" rapidly gained a potent cachet. When Charles Frohman and his associates formed the so-called Trust or Syndicate they made advance national bookings even easier through their careful planning. Frohman was also instrumental in promoting the "star" system, a system fundamentally inimical to stock or repertory. He and other producers discovered that the unique personalities or artistry of some players could fill theatres even when the play itself was weak. Other organizations quickly followed Frohman's practice of affording selected favorites extensive publicity buildups.

Of course, a few performers in musicals had always enjoyed "star" status, even if the word had not yet been coined and their names were not elevated above the title. Mrs. John Wood, John Brougham, Frank Chanfrau, and William Mitchell, although now scarcely remembered, were beloved figures who could often pack houses solely on their reputations. Probably George L. Fox, the era's Humpty Dumpty, was the first "star," in the present sense, of the American musical theatre. Yet for much of the nineteenth century such celebrities were exceptions. As post-war changes spread, however, the lyric stage simply could not resist the trends. So it soon followed that a commandingly beautiful prima donna or a wonderfully engaging clown grew unwilling to be just another member of a "family."

In the nineties, performers began deserting starless troupes to head their own. Yet even here they often kept up the pretense that they were merely first among equals in a unified, ongoing band. Posters proclaimed that the De Wolf Hopper Comic Opera Company was arriving in *Wang* or the Alice Nielsen Comic Opera Company was offering *The Fortune Teller*. And to some extent pretense touched on reality. When Nielsen later toured with a weaker vehicle, *The Singing Girl*, she took along sets and costumes for *The Fortune Teller*, playing it in repertory whenever *The Singing Girl* could not attract sufficient trade.

Even seeming exceptions, notably Lillian Russell, sometimes had to play the game. Miss Russell, the reigning stage beauty of the period, rose to her unique stardom without any apparent loyalty to a particular troupe and several of her early vehicles were mounted independently. But during much of her career she did work with established troupes, making what today we perhaps would call "guest appearances." As such, she was willing to appear in a brief run of one musical while rehearsing a more important one by the same company, or to star in a hastily set-up revival if a major premiere failed. These momentary attachments never lasted long. Her art, her drawing power, and her temperament saw to that. At one point, ascerbic drama critic Alan Dale wryly observed, " 'airy, fairy, Lillian' . . . seems to have settled down to business. She has been singing at the Casino an unusually long time—for her. It must be nearly a year since she has broken a contract."

Other leading ladies often allowed themselves the dubious luxury of a well-publicized tantrum. Tempers and temperaments flared regularly, but they could not mask the fact that these prima donnas were also prima donnas in the best sense of the term. Most had fine voices, excellently trained. Of course, good looks and some personal magnetism were essential. A homely girl or a cardboard performer had no chance. But good looks and personal magnetism alone were insufficient. When Dale published a survey of leading ladies in 1890, he could forget his reservations about Miss Russell's behavior and admit, "She is in admirable voice, and comic opera lovers realize the fact that she is the best singer of her kind that New York has." By reading between the lines even when Dale was critical, one can surmise that other singers possessed reasonably commendable voices. He wrote, for example, of Pauline Hall, "Miss Hall's voice is not at all extraordinary. In fact it is somewhat metallic in quality." Dale stopped short of saying hers was not a good voice, and indeed many other reviewers suggested it was. Ten years later Strang described Christie MacDonald's voice as "sweet, delicate, musical and skilfully controlled," while of the already legendary Miss Russell he gushed, "Her voice, a brilliant soprano, was rich, full and complete, liquid in tone, pure and musical."

In interviews and biographies virtually every prima donna who spoke out professed a burning ambition to sing grand opera. The occasional sorties of troupes to which they belonged into the loftier realms of grander opera seemed only to whet their appetites. Significantly, when Aronson came to write his memoirs in 1913, he attributed what he saw as a recent decline in comic opera singing partly to the lure of vaudeville's higher salaries but primarily to the sudden acceptance of American talent by the Metropolitan Opera and other great opera houses. His implication is clear: the earlier prima donnas of American comic opera were in reality often of a caliber deserving of grand opera.

They were so good, in fact, that they frequently assumed men's roles. Skeptics insisted that these trouser roles, as they were called, simply served as an excuse to reveal a woman's shapely legs, for the parts invariably called for tight-fitting period costumes. The skeptics put their finger on a basic reason for such artificial casting, but the very excellences of these women—their superb voices and exhilarating stage presence—must have also made a strong claim for double duty. No Dennis King or Alfred Drake appeared, whose magnificant singing and compelling acting demanded equal billing. However fine the period's male singers, men such as Tom Karl or William T. Carleton, their acting seems to have been at best mechanical and their personalities lacking in any unique charisma. As a result, they were neglected in this initial rush to musical stardom, and only the great men clowns obtained a share of top billing.

If George L. Fox did not live long enough to move his talents from pantomime and burlesque into operetta, his successors quickly determined where their future resided. Indeed, they came not merely from pantomime and burlesque but from the circus, the variety (early vaudeville) stages, and even from more serious dramatic endeavors. Much of their art remained visual to the end, often dependent on their skills at acrobatics. Combined with their larger-than-life physical movements was their more-ridiculous-than-life physical makeup. Grotesque costumes and grotesque cosmetics were generally the order of the night. As a rule these clowns were unexceptional singers.

Yet within the traditions of their school, there was substantial diversity. Stocky Jefferson De Angelis was often looked upon as the most loveable, and he was one of the pioneers in moving clowns away from the grotesque into the realm of dialect buffoonery. Francis Wilson brought a singular grace of movement to his turns and tumbles and imbued even the lowliest tramp with a discernible aristocracy. Most popular of all was towering De Wolf Hopper, one of the few fine singers among the cutups. He entered with an air of stern authority, but it was an authority quickly and ludicrously punctured. Probably only he was a sufficiently potent draw to deserve his name alone above a title.

In a sense the women were making up for lost time. Francis Wilson recorded that "There were no women on the stage in minstrelsy, and few or none in the audience of 'Variety' because of its coarseness." All this changed on both sides of the footlights as book musicals supplanted the vogue of minstrel shows. There had, of course, been famous ballerinas and a few celebrated comediennes, but, except for the French queens of imported opéra bouffe troupes, prima donnas came into their own on the popular lyric stage only at this juncture. The relatively unobjectionable nature of the stories they acted out and the lyrics they sang prompted other women to come to watch them.

The growth of a star system and a gradual decline of repertory and stock operetta companies, both in the nineties, ineluctably altered the very nature of the era's comic operas. But other changes also determined that the genesis and essence of these musicals would be noticeably different by the turn of the century. For the first decade, importations vastly outnumbered native efforts, and embarrassingly outdistanced them in quality. Curiously, the count of supposedly démodé French opéra bouffes premiering in New York exceeded the total of English, German, and native works, or, for that matter, merely of other importations. But the runs awarded these Parisian confections, with few exceptions, were relatively short. A careful examination of these opéra bouffes disclosed a marked difference from the earlier Offenbachian successes. The often veiled topicality and customary satiric thrusts of the first works had given way to more romantic stories, which were more romantically treated. And the music, while still invigorated and bubbly, reflected the new romanticism as well.

The story Charles Lecocq set to music in the popular *La Fille de Madame Angot* suggests the changes. Not only was there no genuine political satire, covert or otherwise, but a political satirist was sourly portrayed. Madame Angòt's daughter Clairette loves the muckraking songwriter, Ange Pitou, and not the barber she is slated to marry. Knowing that Ange has regularly been arrested for singing his defaming ditties, Clairette contrives to postpone her wedding by publicly singing one of Ange's songs and getting herself jailed. The song she sings attacks Barras, the head of the government, and his mistress, Mademoiselle Lange. Clairette is unaware that Ange has accepted a large bribe from Barras to stop his attacks. Mademoiselle Lange, curious why a young girl would sing about her so scathingly, has Clairette summoned to her chateau. The two women discover they were childhood friends. Clairette learns that Ange is not merely corrupt but has been unfaithful to her. She decides to accept her mother's wisdom and to marry the barber.

With time, English and American comic opera would evince a similar evolution. In fact, a more romantic style of English comic opera, as exemplified by Edward Solomon's works, developed concurrently with Gilbert and Sullivan's marvelous nonsense. The vicar in *The Vicar of Bray* will not hear of his daughter Winifred marrying his priggish and poor curate, Henry Sanford. The mercenary vicar demands she marry Thomas Merton, while the vicar himself proposes to marry Merton's rich widowed mother. Sanford goes off to become a missionary among the cannibals. While he persuades them to abandon their heathen ways, they convince him to abandon his priggishness. He returns not only to wed Winifred, but to become the new Vicar of Bray. Although Sydney Grundy's libretto and lyrics veered noticeably toward the romantic, they remained witty throughout—filled with felicitous jibes at pomposity, religious hypocrisy, and money-grubbing. But much of Gilbert's artificiality had been tempered or had disappeared. What paradoxes and absurdities Grundy dealt with were far closer to nature, so that his characters were far more acceptable as real human beings than Gilbert's. Nor could Solomon's music be dismissed out of hand. It was pleasantly melodic and appropriate. But music, lyrics, and libretto alike were relatively commonplace and thus not nearly so beguiling or memorable as Gilbert and Sullivan's.

In the face of an even then obvious trend toward more romantic operetta, Gilbert and Sullivan countinued to work sturdily in their own traditional mold. Except for their relatively late *The Yeoman of the Guard*, their works remained squarely within the comic opera form they had established. *Patience* and *The Mikado* were the most rapturously received by American audiences. Some small part of *Patience's* success was the coincidental visit to America of Oscar Wilde, who had been so deliciously satirized in Gilbert's libretto. *The Mikado* was recognized instantly for the masterpiece it was, but it benefited as well from a growing vogue for Orientalia. Although it was not the first musical set in the Far East, it unquestionably consolidated the fashion for Japanese and Chinese locales in the theatre.

German operetta, at least as presented in America, missed the high satire of its rivals even if its romantic stories often offered a share of nonsensical puffery. From the start, it trafficked in a frivolous, cavalier pursuit of sexuality that, as we have seen, never seemed to bother moralistic critics as much as the harmless peccadilloes of opéra bouffe.

Musically, French and German works resorted more often to contemporary dance forms (such as Offenbach's can-cans and galops and Strauss's waltzes) and to more immediate folk music. English works, however great, were inclined to wear a "higher" musical art on their sleeves.

The history of the Casino in these years parallels much of operetta's history. The theatre's first season, 1882–83, saw not a single American comic opera offered to Broadway—at the Casino or elsewhere. (The season was, coincidently, the last in which minstrelsy kept a major house lit all year.) During the house's first decade, thirty-four musicals were mounted, mostly German with a few English ones. But not a single American piece graced its stage until 1892.

Though no complete American work was offered, American songs were heard often. In this respect, the Casino again typified the era's attitudes. However aware of audience tastes, however careful its casting and excellent its mountings, the Casino fell in line with lesser theatres and producers in the questionable practice of interpolation. Gilbert and Sullivan quickly developed a reputation that discouraged tampering (and contracts they later awarded may have specifically prohibited it), yet, the Englishmen apart, no foreign composer or librettist was able to protest effectively. In the Casino's very first offering, Strauss's *The Queen's Lace Handkerchief*, at least one comic topical number, "The Dotlet On The I," was added for Francis Wilson. In his autobiography, Aronson contends that the insertion was in no small way responsible for the operetta's success. One can only wonder what Strauss must have thought, if he was aware of it in the first place.

What a later librettist, Edward Paulton, thought of such interpretations has been recorded, by Aronson himself. Paulton's show, *Erminie*, was the most successful in the Casino's early history. Even more vividly than *The Vicar of Bray*, *Erminie* demonstrates the evolution of English comic opera from the comic to the romantic. Its love story was told in a straightforward fashion, the laughs coming almost entirely from the roustabout antics of its two comic villains. The story centered on yet another heroine ordered to marry a man she does not love. In this instance, the comic villains kidnap the bridegroom and take his place. Since they are more interested in thieving than in wedding, the nuptials are delayed. When the groom finally appears he realizes the man the heroine loves is an old friend, and he withdraws in his friend's favor. As Aronson recalled, Paulton told him, "With the antics of some of the people on stage, the many interpolations and its Americanization, so to speak, 'Erminie' will be a fiasco."

Naturally, Aronson's recollections were self-serving, justifying his judgment. That judgment prompted numerous alterations in *Erminie*'s score, brought in from diverse, seemingly disparate sources. For example, "Sunday After Three" was derived from a popular German melody, and "Downy Jail Birds Of A Feather" was culled from Robert Planquette's *Les Voltigeurs de la 32^{e}*. Aronson insisted the latter "fit the situation like a glove." Of course, the expressly written lyric did, but what of stylistic differences between Planquette and Edward Jakobowski, *Erminie*'s German-born composer?

Nor were the high-minded Bostonians above occasional tampering. Barnabee could sometimes be persuaded to embellish an evening with "The Cork Leg," a comic song with which he had been identified in his younger years. And Barnabee was not alone among the Bostonians in adding extraneous material. As he recalled,

> In "Fatinitza," at the risk of upsetting the artistic equilibrium of the harem scene, Mr. Fessenden . . . delighted to spring an irrelevant but stirring song, entitled "My Native Land," words by himself, music by Suppé, though the latter was written for something as far removed from comic opera as oratorio is from ragtime. It was supposed to express an exile's yearning for home-chicken, after a sojourn in Turkey-land.

Perhaps most startlingly, Jessie Bartlett Davis long remembered an instance when she attempted to replace *Robin Hood*'s "Oh, Promise Me" with a new song. She had tired of singing and encoring her great hit (she estimated that at that point she had sung it about five thousand times in two thousand performances), but when she began the interpolated melody she was hooted off the stage and had to return sheepishly singing what Reginald De Koven had written for her. But for the Bostonians, these deviations were remembered precisely because they were exceptions.

In respect to what Barnabee called waving "the star spangled banner of native art," however, the Bostonians were only a year or so ahead of the Casino, and lagged far behind McCaull. As early as 1884, McCaull's Opera Comique Company had courageously mounted John Philip Sousa's *Desiree* in Washington. Like most comic opera troupes the Bostonians did not find a place in their repertoire for domestic efforts until the early nineties. Only with the 1890–91 season did the count of American operettas begin heavily to outnumber importations.

This, then, was basically the theatrical scene when Americans first set about writing comic opera or operetta. (They were to employ both terms regularly, but, except for McCaull, hardly ever used "opéra bouffe.")

3

Early American Attempts at Operetta

Because the English comic operas of Gilbert and Sullivan and those of Edward Solomon, and, to a lesser degree, the earliest Viennese operettas had become the darlings of the playgoing public, Americans had to emulate these sorts of musicals if they hoped to succeed in creating a native musical theatre. The best chance to see one's work professionally mounted lay with a repertory or stock company, if its members could be prevailed upon to move beyond old warhorses and new importations. No doubt the company would want to show its roster to full advantage, so parts had to be created for a full range of good voices, with the understanding that a few "men's" roles would actually be sung by women and some diddling melodies would be inserted for a male comic who sometimes couldn't sing as well as the rest of the cast. Furthermore, reaction against opéra bouffe's salaciousness argued for a clean story and clean lyrics.

Less than a year after *Pinafore*'s premiere, the lively Philadelphia theatre presented a homegrown comic opera, one whose English characters and title plainly announced its debt to the older show. *The First Life Guards at Brighton* was welcomed locally and enjoyed a two-month run, but on the road it was viewed by more jaundiced eyes and soon disappeared. It lasted only long enough to inaugurate what New York derisively labeled "the Philadelphia school" of comic opera.

New York was not any kinder to its own local efforts when they began to appear. Pieces such as *Deseret* in 1880 and *L'Afrique* in 1882 were given the shortest shrift, and deservedly so. In 1882, Washington too mounted a short-lived failure, *The Smugglers*; but this work had the virtue of introducing theatre audiences to a great composer, John Philip Sousa. Sousa's second work two years later, *Desiree*, and his third, two more years on, *Queen of Hearts*, were also unsuccessful.

Practically the only American comic opera of the 1880s to enjoy a long run and reap substantial profits was another example of "the Philadelphia school," Willard Spenser's *The Little Tycoon*. Naïve, banal but curiously likable, this tale of a lover pretending to be "His Royal Highness Sham, The Great Tycoon of Japan" in order to win the approval of his fiancée's stupid, snobbish father played two and a half months in Philadelphia, then toured the country for several years. It carried with it the first song from American comic opera to achieve national popularity, a trite, pleasant waltz called "Love Comes Like A Summer Sigh." However, Spenser was, and chose to remain, a loner. Although several of his later works were also successful, he actually moved only on the periphery of the theatrical world.

It fell, then, to McCaull and, several years later, to the Bostonians to really get the ball rolling. McCaull's motives were no doubt primarily mercenary; but the Bostonians were, as mentioned, genuinely concerned with encouraging native creativity. Interestingly, McCaull and the Bostonians both went to the same writing team, or, more accurately, the same team applied to McCaull and the Bostonians. These collaborators were Harry B. (for Bache) Smith and Reginald De Koven. De Koven, a fine musician, was to compose one supremely beautiful score, then spend a quarter-century unsuccessfully attempting to repeat that achievement. Smith, on the other hand, was a fascinating phenomenon. By his own count he wrote librettos for over three hundred shows and lyrics for approximately six thousand songs. At least one hundred twenty-three of his shows reached Broadway. He was indisputably the most prolific librettist and lyricist in the history of our musical stage. And while he was rarely, if ever, inspired, he was almost always competently pleasant.

Born in Buffalo in 1860, Smith moved with his family to Chicago while he was still in grammar school. Chicago's playhouses witnessed the earliest stirrings of his life-long love affair with the stage. Although his first small paychecks came from publishing houses and newspapers, he dabbled with lyrics and librettos even as a teenager, working mostly with amateur composers in amateur treatricals.

In his autobiography Smith confessed, "Reginald De Koven and I were enthusiastic admirers of the early Gilbert and Sullivan operas and when we started to work on our first piece we decided to bestow on the distinguished collaborators the sincerest flattery of imitation." This flattery not only took the form of studiously copying fundamental patterns, but went so far as to set the action of the story in that exotic jewel of the British empire, India. Quite probably the example of *The Mikado* had suggested so far off a locale. Neither man took note of how many times Gilbert and Sullivan allowed their tales to unfold in contemporary English settings. Nineteenth-century American comic operas set in America or even just in the nineteenth century were rare indeed, and, no doubt, the example of Smith and De Koven's earliest works served as models to other aspiring writers. By leaving contemporary and native material to musical comedy (as well as to revues, which began to appear in the last half of this period), comic opera authors may have unintentionally precipitated a subtle alienation of large segments of their audiences, for even their domestic works were immediately branded as something slightly foreign and unnecessarily farfetched.

The first Smith—De Koven collaboration, *The Begum*, was no more than a modest success. However, since it was neither Smith's best nor worst, it can well stand as typical of his and all other comic opera librettos of the period—those earliest years when comic operas strove to be thoroughly comic. From Gilbert and Sullivan, no doubt, came both the ludicrous names given the characters and the preposterous situations. Like most operettas of the day, it was in two acts, with just a single set for each. (Even when the era's comic operas had three or four acts, they rarely offered more than one scene per act.)

Smith set his first act in the courtyard of the Begum's palace, somewhere in northern India. The court jester, Jhust-Naut, and the court astrologer, Myhnt-Jhuleep, meet and through their dialogue establish the basic situation.

> *Jhust.* But why did you predict Howja-Dhu's death?
>
> *Myhnt.* His son, Pooteh-Wehl, loves my daughter, Aminah; but they cannot marry until he becomes a self-supporting institution. When Howja-Dhu dies, his son becomes Prime Minister; so Pooteh-Wehl induced me to predict his father's demise, which I did. Fortune has favored me in this matter. You know our Begum always marries her General-in-Chief. When she tires of him she declares war, and bids her husband meet a hero's death. He does so, and our Begum marries his successor. Thus she has married nearly a hundred Generals-in-Chief. Isn't that singular?
>
> *Jhust.* No; I should say it was plural. (Chuckles)
>
> *Myhnt.* Don't talk shop. Our Begum is at war now, and—happy thought! perhaps Howja-Dhu has been killed. In that case I should see in all this the footprints of the hand of fate.
>
> *Jhust.* But Howja-Dhu never goes to battle.
>
> *Myhnt.* True; I had forgotten that, though he is an officer of high rank, he leaves all fighting to his substitute, Klahm-Chowdee.

Since the crafty Howja-Dhu, and not the Begum, was Smith's principal figure, audiences knew that whatever absurd complications might arise, the General-in-Chief would be alive and happy at the final curtain. The Begum, in fact, marries Klahm-Chowdee. (America's own commander-in-chief at the time, President Cleveland, was known to have hired a substitute to fight for him during the Civil War, so Howja-Dhu's behavior may have been a sly dig at a prominent government official, much as Sir Joseph Porter's career in *Pinafore* bore a ludicrous resemblance to W. H. Smith's.)

From beginning to end, Smith's humor was light and smiling, deriving not so much from sharp observation or cocky nastiness as from wordplay. His sentences were generally full, and speeches often ran to whole paragraphs. It was a libretto for a patient, decorous age.

Smith's debt to Gilbert is even more obvious in his lyrics (and songs occupied three-quarters of the show). However romantic these comic operas may be deemed, the earlier ones, certainly, allowed relatively little time for romantic love songs. Aminah and Pooteh-Wehl were allotted one in each act. Their first act duet is typical—demure and leaning heavily on poetic apostrophes Victorians so adored.

> Here ye the birds above me singing;
> Listen, blossom, brook and bee;
> Here ye clouds in azure winging,
> That my love loves only me.

More to modern tastes may be the comic love song sung by Myhnt-Jhuleep and his fiancée, Namouna. Its lyric records the same embarrassed moment Alan Jay Lerner handled so deftly seventy years later in the film *Gigi* when Maurice Chevalier and Hermione Gingold sang "I Remember It Well."

> *Nam.* Do you remember, sir, a night
> When we together strolled
> Amid the radiant moonlight
> And love-lorn ditties trolled?
> When you my face and figure praised
> As you alone could do;
> And in my hazel eyes you gazed
> With yours of azure hue?
>
> *Myhnt.* (Reflecting) In vain the hall of memory I scan,
> I think it must have been some other man.

In the next stanza the confusions are reversed. Myhnt-Jhuleep waxes nostalgic over their first meeting, and Namouna must confess her recollections are at odds with his. Perhaps so formal a balance says something about the artificiality of the period's librettos. The disagreements are undeniably comic, but how much more realistic, believable, and therefore poignant is the one-sided situation in *Gigi*, where Chevalier fails to remember accurately any of the details Gingold recalls.

Smith seems to have been far more at ease with patter songs, although nowhere does he approach Gilbert's brilliance. Words flow more naturally in these songs than in many of the serious lyrics, and in them Smith does manage to make some acute observations on human behavior with wit and rhyme. Thus, Howja-Dhu's entrance:

When war began, I said, said I: "My duty is to go."
(They drafted me, you see, and so I could not well say no;)
But as my life was far too precious to be thrown away,
I hired a wretched substitute at rupees twain per day.
"Go forth," said I, "and for your country perish in the fight;"
You'll say 'twas inconsid'rate, and I rather think you're right.
But oh! the moral beauty
Of doing of one's duty,
And oh! the sense of calm content at thought of duty done!
And oh! the satisfaction
Of peaceable inaction,
And oh! the peaceful joy of looking after number one!

"Number one" may seem a curiously contemporary phrase, and Smith obviously chose to insert a few examples of 1887 slang. Late in the play the Begum sings:

What though my griefs cannot be called extensive,
For marriage tends to dissipate "the blues."

Critics would shortly complain that many comic operas employed too much slang.

But an occasional colloquialism could not change the fundamental impression comic opera gave, which instantly separated it from musical comedy. Musical comedy, for all its roustabout absurdities and unreal conventions, for all its garish sets and gaudy costumes, suggested little more than an extension or exaggeration of reality, a suggestion underscored by modern, familiar settings and more or less ordinary characters. Conversely, a well-bred artificiality was the keynote of operetta, at least during this period. If the names of some musical comedy characters were bizarre, names of operetta figures, especially at first, could be even odder.

Furthermore, operetta characters rarely represented the man in the street but rather kings and queens, gods and goddesses, supernatural beings, and other uncommon folk. After a few seasons, as has been said, the more outlandish and satiric elements of operetta gave way to a sentimental romanticism. A slight social leveling sometimes crept in as lovelorn princes and princesses, their hearts surrendered to village maidens and ordinary soldiers, took center stage. But the leveling was more apparent than real, more momentary than enduring. Village maidens and ordinary soldiers lent their voices to soaring last-act finales knowing they would soon be members of royal families and probably have a throne of their own.

Indeed, those soaring last-act finales brought home the truth that nothing separated comic opera from other genres as much as its music. Audiences could leave Ned Harrigan's *Reilly and the Four Hundred* singing, or practically talking, "Maggie Murphy's Home." They could probably pick it up correctly after one or two hearings. But a song like "Oh, Promise Me" from De Koven's *Robin Hood* took careful study.

Four composers dominated the American comic opera in these fin-de-siècle years: Reginald De Koven, Julian Edwards, Victor Herbert, and John Philip Sousa. Undismayed by *The Begum*'s brief existence, De Koven and Smith embarked on a second work, *Don Quixote*. The partners conceived it as a perfect vehicle for McCaull's ensemble, considering the towering, deep-voiced De Wolf Hopper (who had played Howja-Dhu) a natural choice for the Don. But McCaull was under contract to mount several imported operettas, so the new work was offered to and accepted by the Bostonians. What changes, if any, were made to please the New Englanders are unknown. Clearly, there were not enough. The work opened to sharply divided notices. Significantly, Smith felt it was soon dropped from the repertory because the roles were not as comfortably suited to the Bostonians as they might have been to those players in McCaull's group for whom they had been specifically created. Thus, for all practical purposes, De Koven's great masterpiece, *Robin Hood*, was America's introduction to the composer. This time the parts were written with the Bostonians in mind, so everything proceeded smoothly. The season was 1891–92, only the second theatrical year in which native material outnumbered foreign. (Edwards would make a successful debut that same season, Herbert would call attention to himself three seasons later, and Sousa, ignoring his earlier failed attempts, would triumph the following year.)

Smith tampered little with the centuries-old legend of Robin Hood. The Sheriff of Nottingham wrongfully deprives Robert, Earl of

Huntington, of the earl's lands and bestows them instead on Guy of Gisborne. He also bestows on Guy the hand of his ward, Maid Marian. This compounds the earl's injury since he has fallen in love with Marian. Furious, Robert becomes an outlaw and assumes the name of Robin Hood. Through the treachery of one of his band Robin Hood is captured and sentenced to death. But first he is ordered to witness Guy and Marian's marriage. At the same wedding the sheriff plans to forcibly wed the betrothed of another member of Robin Hood's band, Alan-a-Dale. As the wedding begins, Alan-a-Dale, hiding in the bushes, reminds his sweetheart of her vows by singing "Oh, Promise Me." The band rushes in with a pardon from Richard the Lion-Hearted just in time to set matters right.

In keeping with a still lively tradition Smith and De Koven made Alan-a-Dale a trouser role, thereby giving Jessie Bartlett Davis a certain theatrical immortality. Similarly they gave Henry Clay Barnabee a life-long meal ticket by making the sheriff a comic, almost lovable, villain. Each of the great male singers was awarded a memorable musical moment. For Tom Karl's Robin Hood they wrote "An Outlaw's Life"; for W. H. MacDonald's Little John, the opera's second big hit, "Brown October Ale"—second only to "Oh, Promise Me"; and for the Will Scarlet of Eugene Cowles, a superb, deep-throated singer who moved handily from villains to romantic leads to comic parts, they created a lighthearted number, "The Tailor and The Crow," in which Cowles was accompanied by a humming chorus.

Robin Hood offers another instance of writers moving away from the outlandishly comic into more realistic realms. Of course, in *Robin Hood*'s case characters and story were both dictated by long-established legend. Nevertheless, if far-fetched names and situations were ruled out even before Smith and De Koven sat down to write the work, the style they chose indicates they had every intention of creating a more romantic piece. The patter of Smith's earlier lyrics was virtually eliminated; the paradoxical humors gave way to more earthy comedy. All the figures, given the sentiments and stagecraft of the era, were essentially three dimensional and believable. And De Koven's music eschewed flights of comic fancy, settling in large degree for sturdy traditional forms—a move that probably pleased the composer no end.

De Koven was born in Middletown, Connecticut, in 1859, son of a clergyman. When he was thirteen his family moved to England, where he was educated at select private schools and at Oxford. He then studied music in Germany and in France, one of his professors there being Leo Delibes. On returning to America he married the daughter of Illinois Senator Charles B. Farwell. Farwell found his son-in-law a sinecure position in his family business so that the young man could spend his time composing. De Koven seems to have been extremely conscious of his social standing, and while those he allowed to befriend him remembered him as charming and fun-loving, the monocle he occasionally wore, his English accent, and the other airs he affected alienated many stridently democratic associates.

However dangerous it is to read a man's personal makeup into his art, in De Koven's case there seems to be some justice in it. His thorough musical education, his social stance, even his early religious background appear to have influenced his compositions. Certainly no one ever challenged his musicianship. His mastery of the variegated forms of his day was total. In *The Fencing Master* the musical program listed, in the following order, a tarantella, a gavotte, a habanera, a waltz quintet, a barcarole, a marinesca, a serenade, a march, a second serenade, and an aubade, alongside some less specifically identified solos, duets, and choruses. Yet while all the forms were duly observed, their spirit was woefully wanting. The music was bloodless. The melody of the barcarole is just barely pretty, and misses that dreamy magic that sets the listener sailing on moonlit waters. De Koven's melody for the "Gipsy Song" in *The Highwayman*, a work generally accounted his most artful, would almost certainly draw an amused shrug from a true gypsy fiddler-king. Certainly, it cannot stand comparison with the unforgettably impassioned dances and serenades Victor Herbert created one year later for *The Fortune Teller*.

De Koven's musicianship perhaps best manifested itself when he came to write the concerted finales early comic operas offered at the end of every act. His first-act conclusion to *Rob Roy*, the gathering of the clans, is especially stirring. These finales seem as close as De Koven could let himself come to creating real theatrical tension. Otherwise, there is a politeness to De Koven's music, a sense of careful propriety. His songs are always on their best behavior. At times one wonders if he didn't perceive them as intimate society salon pieces rather than large-scale theatrical numbers. No marvel, then, that so many homey sopranos have trilled "Oh, Promise Me" at church weddings, or that until recently, when its vogue has begun to fade, the song was elevated almost to the status of a minor hymn.

De Koven's shortcomings soon told, particularly his inability to create compellingly memorable melodies. He quickly took to composing themes reviewers termed "reminiscent." And although several of his comic operas that followed *Robin Hood* enjoyed long runs, by 1901 one could read between the lines of a rather arch notice in the *Dramatic Mirror* a sense of impending exhaustion:

> Reginald De Koven is among the most prosperous of American opera builders. He secures librettos made in the best jingle-foundries. Around these frameworks he erects musical structures that are, to say the least, pleasing. He selects his bricks from the famous operatic palaces of the past. He lays them skilfully. Occasionally he sets in a gold brick of his own making. Thus Reginald De Koven utilizes second-hand material. . . .

In an interview with the same trade paper five years before, De Koven had spent most of his time boasting of how his orchestrations marked an advance over those of Offenbach and other opéra bouffe composers. Offenbach's orchestrations had been too martial and unsubtle to please the slightly prissy, snobbish De Koven. He attributed an appreciation of his advances to the growth of a discerning leisure class in America. He lamented the confusion among comic opera, operetta, and opéra bouffe, and unsuccessfully attempted to clarify the distinctions. But to charges of repetition and even of outright plagiarism De Koven had little to say, except to insist that all composers were "imitative." This was especially true of American composers, he added, since as a nation we were "too young" to have developed a native musical style. He argued simply, "What we need is music in America, not American music."

In 1903 De Koven assisted the Shuberts in building the Lyric Theatre on 42nd Street in New York, hoping the house would be a home for comic opera. That these comic operas would reflect De

Koven's own style went unsaid, as did the sad truth that both the composer and his style of operetta were all but written out. Little came of the project. Ten years later Barnabee observed:

> And yet Mr. De Koven has tried in vain to find a successor to "Robin Hood." "Rob Roy" might have done it, possibly, if it had been entrusted to the Bostonians first.
>
> The "Red Feather" was to do the trick, "Happy Land" was to fill the bill, the "Student King" with its male chorus was a "sure pop" and the "Golden Butterfly" was to be a gilded winner, but
>
> Mr. De Koven is *still trying*.

That same year, 1913, saw the production of *Her Little Highness*, De Koven's last full score for Broadway. When it failed, De Koven privately acknowledged he had nothing further to say in the musical theatre and retired. He lived until 1920. His fame and that of *Robin Hood* outlived him by several decades. A road company of *Robin Hood* was brought into New York as late as 1944, albeit the revival was shabbily mounted and had to buck the exploding popularity of newer musical forms. But thanks largely to "Oh, Promise Me," a vestige of former renown still clings to De Koven and his masterwork.

Julian Edwards has not been so lucky. It was not his good fortune to write a song that was to be wrenched out of context, removed from its gauze and canvas forests, and passed down lovingly from generation to generation. Yet given the esteem in which he was once held and the success he once enjoyed, his plunge into nearly total oblivion is remarkable. Many critics of his day considered him a finer musician, a more serious, purposeful artist, and a more theatrically attuned melodist than either De Koven or Sousa.

Edwards was born in England in 1855. He studied music in both Edinburgh and London before accepting a post as conductor with the Carl Rosa Opera Company. It was as a conductor that American producer James C. Duff persuaded him to emigrate to America.

Edwards's first musical, *Jupiter*, was written for Duff's company and premiered in June 1892, just eight months after *Robin Hood*'s New York premiere. His librettist for the occasion was Harry B. Smith. Smith's tale tilted back a bit toward the more artificial comic plots of the preceding decade. It told how Jupiter comes to earth disguised as a shoemaker in order to seduce the shoemaker's wife. Edwards's score attracted little attention. Digby Bell's comic antics walked off with the critics' praise, and such was the power of leading comedians that Bell was able to interpolate his own rendition of "Annie Rooney." (Such an interpolation suggests not only a lack of feeling for tone in the show, but that Bell, for all his clowning, could in his turn tilt the emphasis away from Smith's attempt to restore something of a pervasive comic spirit.)

Edwards's second score was a different matter. For *King Rene's Daughter* Edwards went to the same Henrik Hertz drama that was the source of Tchaikovsky's *Iolantha*. The story was pure romance—the love of a young count for a blind princess. Plans originally called for *King Rene's Daughter* to be presented on a double bill with Gounod's *Philemon and Baucis*—not at an opera house, but at a regular Broadway theatre. A prima donna's illness delayed Edwards's premiere, and Gounod's piece played the first week unaccompanied. When Edwards's work did raise its curtain, it dramatically split the critics. The *Herald* protested that the music "wallows in Wagner," and several other papers insisted both works belonged more properly up the street at the Metropolitan Opera. But the *Dramatic Mirror* typified the aye-sayers when it praised the score as "musicianly and musical." The public would not accept the bill, so it was withdrawn after two weeks.

But praised or denounced, Edwards was marked as a composer to be reckoned with. Except for a Lillian Russell vehicle, *The Goddess of Truth*, and a later work called *Princess Chic*, Edwards followed *King Rene's Daughter* with a series of musicals that Broadway and the road or at least the road joyfully welcomed: *Madelaine* (with its charming waltzes), *Brian Boru* (although here some critics lamented the absence of local color in the music), *The Wedding Day*, *The Jolly Musketeers*, and *Dolly Varden*.

Edwards's greatest success was his 1902 *When Johnny Comes Marching Home*, written to a libretto by his favorite collaborator, Stanislaus Stange. However stilted and naïve some of Stange's dialogue may now appear, the comic opera is certainly one of those neglected gems that deserves a rehearing. Both the story and the music that accompanied it testify to the inexorably evolving art of comic opera. The preposterous events and figures that enlivened every turn in the works of the older school were increasingly left behind. Situations, characters, and even characters' names were kept well within the bounds of ordinary dramatic license. No one need dwell soberly on the gleeful head-choppings of previous comic operas, but, despite its sing-song gaiety and its happy ending, a work such as *When Johnny Comes Marching Home* moved along to genuinely somber underchordings. Comic opera was continuing slowly but surely to become less pervasively comic.

One of the earliest comic operas set squarely in America, *When Johnny Comes Marching Home* went to the Civil War for its tale. A Southern belle, Kate Pemberton, loves a Union colonel, John Graham, but she is not above helping her Confederate cause by stealing Northern military secrets. Graham discovers what she has done, but understands her divided loyalties. Just a generation after all the embittering carnage, the musical bravely attempted to portray both sides sympathetically, although, of course, its most eloquent plea extolled the glories of the Union. That moment came when the hero led the chorus in a rousing anthem, "My Own United States." The song's melody was new to first-nighters, but its hymnlike fervor was not. From the overture's first notes to the finale's last, the show's music was of a sort its audiences could feel instantly at home with. Edwards's sensitive artistry saw to that. Aubades and barcaroles and tarantellas were not part of America's real musical idiom, and so would have been out of place.

But, in fact, these foreign forms were rapidly disappearing even from American operettas set overseas. Authors and producers had discerned that American audiences responded more immediately to homier, simpler, and more familiar musical patterns. Because of its very nature comic opera would never be able to fully abandon its older, continental inspirations, but it was to employ them more cautiously. To the extent that these forms sounded alien to Americans, their disappearance from musical programs represented another move away from extreme artificiality.

Edwards sidled closer to a stagy verisimilitude by another means. Beginning with strains of "When Johnny Comes Marching Home" at the start of the overture, Edwards cleverly interlaced his own songs and orchestrations with historic favorites. His first act finale moves from his own "I Could Waltz On Forever" through to "My Own United States" with concerted references to "Tramp, Tramp, Tramp," "The Battle Hymn Of The Republic," "Marching Through Georgia," and "Dixie." "Old Folks At Home" was sung as a countermelody at one point in the show.

When Johnny Comes Marching Home's weakest musical moments were probably its love songs. While not as constricted as De Koven's, and redeemed by a certain gentle sweetness, they still fell far short of suggesting or provoking passion. Possibly a correct sense of period prohibited Edwards from including one of those free-swirling waltzes that had made *Madelaine* so attractive.

Some of the work's most charming moments come with what must even then have begun to seem period pieces, echoes of minstrelsy such as "Katie, My Southern Rose," "My Honeysuckle Girl," or "Sir Frog And Mistress Toad" (this last, one of those anthropomorphic ditties that had been regaling playgoers at least since the days of opéra bouffe).

For all his artistic integrity, Edwards, like De Koven, lacked that free-flowing well of great melody that separates the mere musician from the master. And like De Koven, Edwards was always to find Victor Herbert ready to demonstrate that difference. One of the most pleasant songs in *When Johnny Comes Marching Home* is "Fairyland." Yet it pales beside "Toyland," which Herbert introduced a year later.

But 1902 audiences could not and need not make such comparisons. They judged what they saw and heard, and they liked it. Received with applause almost everywhere, *When Johnny Comes Marching Home* was kept on the road for several seasons and given a major revival as late as 1917. Its success and the praise heaped on Edwards's work no doubt weighed in Mme. Ernestine Schumann-Heink's considerations when she elected to star in his 1904 comic opera *Love's Lottery*. But Edwards's powers failed him, and critics branded his score as turgid.

Unfortunately the failure seems to have been lasting. Perhaps because the composer sensed his inspiration waning, Edwards made a peculiar change. Just as Franz Lehar's *The Merry Widow* revitalized operetta in 1907, Edwards abandoned the genre and tried his hand at musical comedy. When three of his shows were presented in the 1909–10 season, none with exceptional success, Edwards began to withdraw from the theatrical arena. Yet such was his reputation that Broadway impresario Robert Grau could conclude in that same season, "Julian Edwards writes opera of a more classical character, and has a list of successes to his credit that will compare most favorably with some of the European celebrities who have obtained large royalties in all quarters of the globe." But for all the fame and respect once Edwards's, probably not half a dozen theatre lovers could whistle a single of his themes today.

Victor Herbert is another matter, although even his fame and familiarity seem to be receding in the last few years, with ever increasing speed. Revivals and reevaluations could change all that, and if operetta can ever throw off the unfortunate reputation it has carried in recent decades, Herbert should almost certainly regain the respect he deserves as our finest composer of turn-of-the-century operetta. Indeed, it can be argued that he was the first in the great trinity of towering masters our musical stage produced, a direct line from 1894 to the present by way of Herbert, Jerome Kern, and Richard Rodgers.

Without modesty, Herbert might well have questioned being placed among our finest operetta composers. For if, by modern lights, all of Herbert's works are operettas, the composer himself, at least in interviews during his last years, insisted many of his shows were musical comedies, definitely not comic operas. Since several of the shows he thought were musical comedies were nevertheless listed as operettas in their original advertisements and vocal scores, his arguments may have been prompted to some extent by an understandable desire to be stylish. Throughout Herbert's active career, comic opera and musical comedy vied to win fashion's fickle blessing. For years that nod alternated almost whimsically between the two schools, and in Herbert's final years seemed likely to ultimately settle for musical comedy.

At the time Herbert appeared on Broadway the first great era of comic opera was coming to an end. Gilbert and Sullivan, Solomon, von Suppé, Strauss, and Millöcker either had died or had written themselves out. While Americans were at last mounting largely American material, except for *Robin Hood* and possibly one or two other De Koven works, little was offered that seemed likely to hold playgoers' affections. At the same time, the great comic opera repertory and stock companies were faltering (McCaull had died a week before Herbert's first work premiered). The musical stage was in flux.

Still, most of Herbert's early works (before he retired from the lyric stage for three years to head the Pittsburgh Symphony), were written either for the Bostonians or for two breakaway companies. The first of these ensembles was led by ambitious, fiery Alice Nielsen, and she took with her several of the Bostonians' finest singers. Indeed, as mentioned earlier, it was for the Alice Nielsen Comic Opera Company that Herbert composed his first masterpiece, *The Fortune Teller*.

Dramatically, there could be no doubt that this operetta was conceived as a star vehicle, not as a work for a genuine ensemble. Miss Nielsen was given not one but two principal roles to play, those of two look-alikes, the saucy yet innocent ballet student, Irma (who doesn't realize she is an heiress), and the knowing, darker gypsy, Musette. In order to rid herself of a pestering count and allow a handsome hussar to court her, Irma persuades Musette to take her place when the count comes to visit her. Complications follow when Musette's gypsy lover, Sandor, appears. If the librettist, our old friend Harry B. Smith, catered to Miss Nielsen in his story, Herbert demonstrated that he retained some allegiance to the older idea of writing for an ensemble. True, he gave Miss Nielsen the coy "Always Do As People Say You Should" as well as a great gypsy number, "Romany Life" ("Czardas"), the latter sung with the chorus. However, the hussar was hardly ignored. He was awarded the stirring "Hungaria's Hussars" and, with the heroine's rival, the lovely waltz-duet, "Only In The Play." For Eugene Cowle's Sandor, Herbert composed the rousing "Ho! Ye Townsmen," the lively "Gypsy Jan," and, best of all, the operetta's most beloved and enduring number, "Gypsy Love Song." Herbert almost certainly had in mind that some ongoing ensemble would want to use *The Fortune Teller* when Miss Nielsen was finished with it.

Herbert's second breakaway afforded a better glimpse into the future of operetta. The Bostonians' manager, Kirk LaShelle, formed what he called the Frank Daniels Comic Opera Company to present vehicles created expressly for the genial little comedian. In truth, the Daniels Comic Opera Company was nothing of the sort. Only a few minor performers moved with Daniels from show to show. Using the guise of a traditional ensemble LaShelle was simply mounting independent productions for a single star. Indeed, when Daniels soon left LaShelle to perform for young Charles Dillingham, all pretense of an ongoing comic opera troupe was abandoned. With Daniels the cynosure and with his supporting players cast more for their comic ability than for their vocal artistry, Herbert necessarily had to compose in a more restricted manner. Tessituras became more limited, emotions played with more on the surface.

If Herbert's obvious artistry could still be displayed in brilliant, brief orchestral passages, the songs had to have a more mundane quality. The rose-colored, high romanticism of his comic opera material would have seemed out of place in a story in which a home-spun, apple-pie American confronts ludicrous obstacles—even if those obstacles are encountered in some far-off land. Such was the situation in *The Idol's Eye* and in later vehicles such as *The Red Mill*, *It Happened in Nordland*, and *Old Dutch*. Even when the comic hero purported to be a foreigner, as in *The Wizard of the Nile* and *The Ameer*, audiences might well have suspected he was a misplaced American. Somehow that willing suspension of disbelief that permitted kings, gods, and the other personages of comic opera to have magnificent voices did not apply here. As a result, one could almost talk the lighthearted lyrics of "The Streets Of New York," but one would hardly dare talk "Ah, Sweet Mystery Of Life."

Even if Herbert's distinction between musical comedy and operetta may have been partly ex post facto rationalization, and even if it today seems somewhat academic, it nonetheless demonstrates the subtle, careful perceptions Herbert brought to his work. And he brought his thoughtful artistry to bear from his very first comic opera, *Prince Ananias*, although the show did not have an auspicious debut.

The operetta may or may not have been written for the Bostonians. Herbert's contract with the players called for him to deliver the score by December 1894, but the work was on the boards in mid-

November, leading the composer's biographer, Edward N. Waters, to conclude that much of the music must have been written before contracts had been signed the preceding June. Barnabee's casual remark that "the music of 'Prince Ananias' was so good that we took it as a running mate with our perennial winner, 'Robin Hood' " appears to confirm the supposition. Since Herbert had met and befriended another Bostonian stalwart W. H. MacDonald in the spring of 1894, the comic opera was probably conceived with the possibility the Bostonians would option it. Certainly an important part was developed for each of the principal members of the troupe.

Prince Ananias dealt with a not dissimilar troupe of wandering players ordered, on pain of death, to dispel a king's depression. It was an inept and inane libretto that clouded appreciation of the music's worth. This was to be the case with many Herbert shows. Apart from Henry Blossom and, some would argue, Harry B. Smith, the composer would never find a really first-class librettist; but unexciting librettos were accepted with more resignation than they are today. Most critics found something kind to say about *Prince Ananias*'s score, and Herbert's orchestrations elicited the highest praise. Reviewers heard in them "real operatic characterization which appropriately supported the leading roles." (Over fifty years after Herbert's death another great orchestrator, Robert Russell Bennett, could still insist that Herbert was the finest orchestrator our theatre had ever known.) So while Herbert's abundant melodic gifts were not yet fully evident, his orchestral material brought drama and its share of musical art to the show.

Herbert's melodic genius was not totally absent. Waters notes,

> The gem of the opera, and one of the gems of all light opera, is "Amaryllis," at the beginning of Act II. The wistful melancholy of Idalia's song, the delicate accompaniment and the interpolation of a dainty minuet make this a truly extraordinary selection. Herbert never attempted anything quite like it afterward, and he may have felt it was too fragile a thing for the operetta stage.

Waters's surmise is assuredly correct, for along with magnificent orchestrations and, eventually, unforgettable melodies, Herbert brought to his works an acute sense of theatrical appropriateness. Unlike De

Koven or Edwards, Herbert would never write a gypsy air that lacked pulsating vitality or wrenching poignancy, or a Celtic melody without some singular Irish grace; he would reject love songs lacking incandescent passions, or bantering ditties devoid of some toe-tapping charm. Herbert's airiest melodies have emotional or dramatic tension.

Herbert's virtues may have reflected his background as much as De Koven's shortcomings reflected his. Herbert was born in Ireland in 1859, but spent important formative years with his widowed mother at the English home of his grandfather, popular Victorian novelist and sometime composer Samuel Lover. His mother started his musical training early, and when she remarried and moved to Stuttgart, she enrolled him at the local conservatory to study cello. After graduation he was accepted as a cellist by the Stuttgart Opera. There he met and married the company's leading soprano. When she was awarded a contract by the Metropolitan Opera, she insisted Herbert be brought over as a cellist for the house. To supplement their income Herbert took to composing and conducting, rising quickly to the post of bandmaster with the Twenty-Second Regiment Band. In short, Herbert grew up not only in a musical home, but a home in which the art of effective story-telling was undoubtedly a subject of fervent debate.

When the theoretical side of Herbert's musical education was finished, he embarked on a more practical one. He had a sharp, active mind, and playing in a pit orchestra must have taught him much not merely about his own instrument but about its effective relationship to the orchestra's other instruments. Moving on from there, he could gauge the relation of the music to the action onstage, and the interaction of the music drama as a whole with the audience. When he wielded his baton before his military band, the compressed theatricality of band music must have provided further insights. And then, of course, his wife, as a superior singer, could offer him additional guidance in writing for the human voice. Consciously or intuitively, Herbert absorbed and manipulated all these intangible elements, extracting from them the often unarticulated principles which shaped his greatness.

His fellow composers acknowledged his supremacy, even if they sometimes attached gratuitous excuses to explain away their own deficiencies. One competitor wrote:

> To talk of light opera is, inevitably, to talk of Victor Herbert. It is always a pleasure to me to include many of his compositions on my programmes. I was handicapped in my own opera-writing by the difficulty of getting first-class librettos and by the arduous demands of my long band tours, but I followed with interest the creative work of Herbert. He was the best-equipped man of his time for this work; curiously enough he always did his best when he was composing for some particular star; evidently it was necessary to him to have a definite human picture in his mind. Herbert was an excellent example of fine musicianship and it was a distinct aid to his success that he could give so much time to his operatic writing. Perhaps no two men in the profession have been paired more often in the minds of the people than Herbert and myself. When I planned to go to Europe with my band for its first overseas tour (which the approach of the Spanish War prevented) Herbert, then conductor of the Twenty-Second Regiment Band, formerly Gilmore's famous group, took my place at Manhattan Beach. . . . In those good old days, the three names most familiar to patrons of light opera were certainly Herbert, De Koven and Sousa. Herbert, I am sure, wrote more light operas of high quality than any other composer in Europe or America.

This autobiographical recollection was, of course, Sousa's. In his own day John Philip Sousa was even more clamorously celebrated and widely adored than Herbert. Tours with his renowned band brought his music and his fame to all corners of the Western world. In country after country he was feted and honored and hailed—as "The March King." Yet among the admiring throngs, there were precious few who knew their idol had also composed hundreds of songs and fifteen operettas. While most of Sousa's marches were played everywhere and played repeatedly, most of his operettas either remained in manuscript or were performed briefly, only to disappear forever. Nevertheless, several of these operettas did succeed and are of exceptional merit. One, *El Capitan*, has in recent decades elbowed De Koven's *Robin Hood* off revival stages, boding well to be the most enduring American comic opera of the nineteenth century. (Remember that all of Herbert's masterpieces except *The Fortune Teller* are twentieth-century works, wherever their inspiration may lie.)

Alone among the giants of early American comic opera, Sousa was both American-born and American-trained, although from the start his background exposed him to foreign influences. His father was born in Spain to Portuguese parents and fled to America during an uprising. His mother was Bavarian-born. Sousa himself was born in Washington, D.C., in 1854. (De Koven, Edwards, Herbert, and Sousa, then, were born within four years of each other, although some students dispute De Koven's birth year, placing it in 1861.)

At the age of six Sousa was given solfeggio lessons, a method by which students are taught to read music by mastering the do-re-mi-fa-sol-la-ti-do scale. A few years later the boy was enrolled at a local conservatory where for four years he studied voice, violin, piano, flute, cornet, and alto horn. His father supplemented his lessons by teaching him the trombone. After running away from home when he was thirteen, Sousa was enrolled in the U.S. Marine Corps band, and continued his musical studies under its director, George Felix Benkert. At nineteen he left the marines to seek employment as a violinist with a theatre pit orchestra. He quickly rose to conductor. In ensuing years he moved from theatre to theatre, not only conducting but composing occasional pieces when necessary and eventurally orchestrating other composers' materials.

Simon Hassler's offer of a position in the first violin section with the International Exhibition Orchestra, the official orchestra of the 1876 International Centennial Exposition in Philadelphia, induced him to turn his back on the theatre for that long holiday summer. But even in the open air of Fairmount Park, the theatre and its music were never far away. Certainly not for the last two weeks of June, when the orchestra's concerts were led by one of its most famous guest conductors, Jacques Offenbach. After the fair closed, Sousa remained in Philadelphia, first as violinist and then as conductor of the Chestnut Street Theatre orchestra.

When *Pinafore* exploded on the American scene, Sousa conducted one of its numerous companies, using his orchestrations in the absence of Sullivan's. At the same time, he and an old Washington friend, Wilson J. Vance, tried their hands at creating an American operetta. Their *Katherine*, though apparently completed, was never produced. However noteworthy so early a plunge into comic opera, the work seems to have scarcely affected Sousa's musical attitudes. But Sousa discovered Sullivan's orchestrations and instantly recognized their superiority. From then on, Sousa placed Sullivan and Offenbach side by side in his private pantheon, and they would provide his melodic and orchestral models when he was ready to embark on his great works.

Of course, Sousa's own martial proclivities gave his orchestrations a unique, identifiable quality. In later writings he would also credit Wagner and Herbert for influencing him. One artist who certainly did not was De Koven. De Koven's "advanced" orchestrations were not to Sousa's taste, although he was too much the gentleman to say so publicly.

Thus, Sousa's musical schooling, the inherent theatricality of band music, and his actual theatre experience certainly fitted him for the task of comic opera writing, much as a similar background prepared Herbert. Why, then, did Sousa write so few comic operas in his long life, why did fewer still reach the stage, and why didn't they burst with melodies that would endear them to the public and plead for their revival? These questions can only be answered equivocally. Far more than either De Koven or Edwards but to a lesser degree than Herbert, Sousa did possess the unique ability to provide his show pieces with a requisite tension. The listener genuinely cares to learn what phrase or invention comes next, and, equally important, is irresistibly caught up in the emotional demands of the song—the thrill of an excited march, the tenderness of a romantic ballad. Yet, this requisite tension also sometimes gives way to an unfortunate tightness, a certain circumscribed formality that asserts itself at the expense of melodic freedom.

Sousa, like De Koven, was bedeviled to some extent by musical art forms his age dictated belonged in comic opera. Herbert's genius allowed him to ignore or override them, but Sousa fought free only on occasion. Still, he was a better melodist than the rarity of revivals of his shows would suggest, so in Sousa's case an ironic inversion may have occurred. Other composers have fallen by the wayside because, as mentioned before, their melodic invention is so small as to preclude revival; Sousa, to no small extent, remains undervalued as a melodist because his shows have not been sufficiently revived. Admittedly, his love songs subscribed to the formality of his day, and so now place a slight barrier to a thoroughgoing modern empathy. Yet surely even a modern hearer cannot help be charmed by "Sweetheart, I'm Waiting." Similarly, a nonsense song such as "A Typical Tune Of Zanzibar," one of Sousa's personal favorites, retains its slightly lunatic delights. And then of course there are the irresistible marches, best exemplified by two still famous pieces, "El Capitan's Song" and *The Bride Elect*'s "Unchain The Dogs Of War."

Sousa did possess an appreciable melodic gift, but he seems to have had a curious blind spot in regard to his operetta songs. His autobiography suggests he may have been relatively indifferent to comic opera. He gives his own works exceedingly short shrift, never really getting down to discussing them musically and viewing them, seemingly, as cues for gossip. For example, the sole reason he offers for *The Smugglers*'s failure is a cloddish performance by one of its principals. Several of his operettas pass totally unmentioned. And throughout his life Sousa never seems to have fought to keep his works before the footlights.

Sousa's biographer, Paul E. Bierley, hints at a reason for this indifference. Despite his vocal studies, Sousa was not very interested in the human voice. In 1899 when he was asked to look back and select the greatest musical artists of the nineteenth century, his long list included composers and instrumental performers but not a single vocal artist. Elsewhere he once proclaimed his favorite musical instrument was the drum—an effective instrument in band concerts but hardly an important asset in comic opera. Singers, in return, often accused Sousa of ignoring human limitations, writing uncomfortable jumps and ranging beyond normal tessituras. Moreover, they complained that his orchestrations were sometimes too brassy and percussive, drowning out their most stentorian efforts.

Sousa also may have created problems for himself in his choice of librettists. This was a common problem of his day, and it must be remembered that time has not been kind to many librettos that were welcomed in their own era. Indeed, Sousa in later years insisted that no finer libretto had ever been written than the one Charles Klein gave him for *El Capitan*. But hailed as it was at its premiere, it was recognized even then that much of its strength derived from the performance of De Wolf Hopper, at once a consummate comedian and a superb deep-throated singer. Since so many of Hopper's best touches were physical, we can only surmise what additional strengths he imparted to the entertainment. But even the recent Goodspeed Opera House revival in Connecticut demonstrated that the book and lyrics, while unexceptional, are sturdy enough to endure.

Klein had devised a genuinely funny story with enough comic mix-ups to compensate for a lack of any overriding wit in his lines. El Capitan was the leader of a rebel band dedicated to the overthrow of Don Medigua, the viceroy of Peru. Medigua captures his nemesis and secretly executes him. He then disguises himself as El Capitan and leads the rebels around and around in circles until they are too exhausted to fight. In the interval, a series of mistaken identities provides much of the fun. But the book could not have survived without a superior score. Sousa and Tom Frost wrote lyrics that were no better than serviceable.

Still, even accepting these cavils as legitimate in no way detracts from Sousa's accomplishments in the theatre. De Koven may have captured the esteem and affection of his contemporaries, but his star has deservedly faded. In the long run, especially if his better works are brought off the shelf, Sousa will probably be accepted as Herbert's only genuine rival, although *Robin Hood* will hold a high place in our early comic opera history.

In an age when only twenty or thirty musicals were presented on Broadway each year, when some of these were either musical comedies or revues, and when most comic operas were written by either celebrated foreigners or the American giants of the day, room for competitors was scarce. And clearly these competitors never threatened to become rivals. A few, notably Willard Spenser, thrived away from New York; a few others, men such as Thomas P. Thorne or Henry Waller, were heard once or twice and then vanished. The world of comic opera composers was small and tightly knit, and one can only surmise and hope that no genius was denied his place in the spotlight. In any case, no great new figure appeared for a number of years. Perhaps partly because of this, operetta fell into relative disfavor, while new forms enjoyed their moment of triumph.

In April 1896 the *Dramatic Mirror* observed, "From present indication it looks as if comic opera will not flourish as flamboyantly as usual this summer. . . . Managers are apparently afraid of competition with roof-gardens." Roof gardens sought the breeziest possible diversions, so musical comedies became the order of those hot summer nights, along with revues, a new form whose vogue the Casino had initiated only two years earlier. And when cooler weather came, musical comedies and revues lingered on, crowding out operettas. Not entirely, of course. Two hits of the 1896–97 season were Edmond Audran's *La Poupée* and De Koven's *The Highwayman*. Nonetheless, comic opera's popularity had waned decisively.

Why did comic opera suddenly take a back seat? No single answer presents itself. Instead a number of reasons may have contributed simultaneously. In "The Decline of Comic Opera" in the January 1905 issue of the *International Quarterly*, W. J. Henderson ties the drop in quality to the demise of the old repertory and stock troupes. Their structure, their sense of continuity and consistency, their ability to keep a flawed show on the boards while its authors improved it, their experience, and, perhaps arguably, their good taste provided the best possible field for creativity. Time has shot this wistful theory full of holes. A composer such as Herbert, who composed masterfully for the old entourages, composed every bit as well for the newer mode. A broader view of history suggests that most artistic flowerings wear themselves out after a score or so years, and this first flowering of operetta was no exception. Indeed, a candid confession would acknowledge that most of the domestic comic operas that dominated American stages of the nineties were patently inferior to the best English and Viennese triumphs of the eighties, as well as to the major opéra bouffe gems of the preceding years. But that same broad view would suggest that early flowerings often produce the seeds of later ones. Operetta's first flowering was to prove no exception.

4

Interregnum: The Withering of Early Operetta and the Rise of Musical Comedy

America at the end of the nineteenth and the beginning of the twentieth century was coming to realize it was a great world power. Innumerable technological achievements and the promise of greater ones ahead engendered a pervasive optimism and a somewhat chauvinistic self-satisfaction. A catch phrase of the era was "up to date." Whatever was new and modern, especially if it was Yankee-contrived, was eagerly embraced. Thus, as the quality and vogue of early comic opera dwindled, Americans turned for light diversion to musical comedies and revues, both so "up to date" and generally topical. Even as staunch a comic opera loyalist as Barnabee proclaimed, "I thoroughly believe musical comedy to be the entertainment of the future." Indeed, as a prophet Barnabee jumped decades into the future when he continued, "In my opinion the story should be more prominent than the music. The latter ought to be written for the libretto rather than vice versa."

No one, of course, claimed that musical comedy or revue were American inventions. Revue's Parisian origins were customarily acknowledged, although a few Irish-American banners were waved for

John Brougham. In any case, these hodgepodges of skits and songs had swept across many nations' stages. Musical comedy's beginnings were harder to pinpoint. Nor was the genre truly new. *Adonis* and *A Trip to Chinatown*, Broadway's longest-running shows at the time, were both fundamentally musical comedies. *Adonis* and its handsome star, Henry E. Dixey, had swept Broadway off its feet in 1884. Dixey played a Greek statue brought to life. He cannot win the girl of his heart, but a bevy of other females will give him no rest. In the end he elects to return to stone. Such preposterous nonsense was like the basic material employed in early comic opera, but both *Adonis*'s words and music were largely vernacular, most of the songs meant to be "talked" rather than actually sung. Charles Hoyt's *A Trip to Chinatown* was more realistic and slapdash. It story recounted the misadventures of some readily recognizable middle-class types out for a night on the town. It relied for laughs more on running jokes, pratfalls, and stock comic situations than on genuine wit. Its music was strictly Tin Pan Alley (with such still-sung melodies as "Reuben And Cynthia," "The Bowery," and a late interpolation, "After The Ball"). It was in every respect typical of the musical comedies Hoyt offered in the nineties.

But in America burgeoning musical comedies and revues could tap a refreshing native spring, one which could impart a distinctly red-white-and-blue coloring to musicals and one which comic opera rightly should not touch. That singular, invigorating American ingredient was ragtime. And ragtime brought with it the promise of a totally new, totally native musical idiom. Virtually from the moment of its first commercial New York hearing in February 1896, ragtime was eagerly welcomed on musical stages.

While show after show quickly added a ragtime tune or two, thus distinctly coloring the fabric of musical comedy, it was the English who once again decreed the basic shape of things to come on American boards. A year and half before Ben Harney introduced ragtime at the Union Square, George Edwardes brought over *A Gaiety Girl*, giving Americans their first taste of what became known as Gaiety Theatre musicals, after the London house where so many of them premiered. These English offerings immediately set new standards for musical comedy writing and sealed the doom of heretofore popular musical comedy styles such as Harrigan's and Hoyt's. Harrigan's roughhouse antics and Hoyt's slapdash joys suddenly seemed primitive and gauche.

For their day the Gaiety shows had well-plotted librettos and light, tripping songs that fit well into the story. *A Gaiety Girl* blended the vernacular appeal of musical comedy with the integrity of the best comic opera. But its vernacular was not that of Harrigan's lower classes or Hoyt's upstarts. Rather, it turned to the more or less polished, self-assured British leisure class. And it substituted style for stylization. Indeed, the Gaiety shows' chief merit may have been a thoroughgoing sense of style, for their comedy was unexceptional and the music not far removed from the music hall. English performers brought over for these shows gave them an added cachet. Mountings were as elegant as the material, so much so in fact that costumes for the shows soon set fashion trends. *A Gaiety Girl*'s basic plot centered on a tale of love leveling ranks, like many a comic opera story. In this instance, a chorus girl at the Gaiety wins the hand of a handsome officer, despite the machinations of a society belle. In a sense, the employment of this motif here presaged the Cinderella era of musical comedy twenty years hence. But instead of obscure peasant girls wedding unlikely kings or princes, working girls wed rich members of the middle class. Other Gaiety musicals, such as *In Town* and *The Shop Girl*, followed, frequently recounting much the same story. Yet, as with comic opera, almost a decade was to elapse before Americans began pouring out memorable musical comedies.

America's first musical comedy genius, George M. Cohan, offered Broadway his maiden effort, *The Governor's Son*, in 1901, and hit full stride three years later with *Little Johnny Jones*. In 1906 he had two popular offerings, *Forty-Five Minutes from Broadway* and *George Washington, Jr.* At about the same time black musicals such as *In Dahomey* and *Bandana Land* began appearing on Broadway, while in Chicago, Joe Howard, Will Hough, and Frank Adams inaugurated a series of melodic, cheery, and wholly American musical comedies. These began with *His Highness the Bey* and continued with *The Umpire; The Time, the Place, and the Girl;* and *A Stubborn Cinderella*.

Although all these shows were generally better plotted than American musicals had been before the arrival of the Gaiety shows, they were less artful or at least less elegant than their overseas competition. Their librettos relied more heavily on vaudeville jokes, and even their songs smacked more obviously of the music hall.

Of course such failings went unnoticed when an audience was confronted by Cohan's electrifying flagwaving in *Little Johnny Jones*'s "The Yankee Doodle Boy" and "Give My Regards To Broadway" and *George Washington, Jr.*'s "You're A Grand Old Flag"; and Howard's down-home sentimentality in *The Prince of Tonight*'s "I Wonder Who's Kissing Her Now" and *The Time, the Place, and the Girl*'s "Waning Honeymoon." Cohan's sentimental side was best revealed by his songs for *Forty-Five Minutes from Broadway*—the title number, "So Long, Mary," and "Mary's A Grand Old Name." But apart from Cohan's best offerings, and, to a markedly lesser extent, Howard's melodies, the music in these shows was not insistently memorable.

Victor Herbert's musical comedies were necessarily exceptions. When circumstances demanded, Herbert modified his lyricism, although since he could enlist the best singers, his modifications were not drastic. Moreover, he could write more restricted material for his comics and more open songs for his lovers. His best musical comedy from this period, *The Red Mill*, is typical. Its plot was one common to its era, the preposterous adventures of two Americans in a foreign land. Kid Conner and Con Kidder arrive, broke, in a small Dutch town ruled by a stern burgomaster. They wangle free food and lodging by helping the burgomaster's daughter escape from the mill where her father imprisoned her when she refused to marry a man of his choosing. For popular comedians Dave Montgomery and Fred Stone, who played Kid and Con, Herbert's songs stayed within reasonable ranges. For example, "The Streets Of New York," written in B flat, moved from the *d* just below the staff to the *f* at the top of the staff, and even here Herbert suggested the *d* within the staff for singers who might find the *f* uncomfortably high. On the other hand, songs given the better-voiced lovers, such as "Moonbeams" and "The Isle Of Our Dreams," ranged more broadly. The differences were not extreme, but they demonstrated Herbert's awareness and skill. In more thoroughly lyric works Herbert allowed himself greater freedom, although even in these he carefully controlled the songs given comedians.

As a rule, however, composers of these newer, lighter entertainments lacked the formal musical training of the comic opera composers. Most were what critics of the time disparagingly branded "one-finger" composers, adequate melodists who often could not write a simple piano part, let alone orchestrate a work. Some were also performers of great personal magnetism but little vocal ability who primarily were anxious to provide themselves with material they could practically talk, or, again as the critics of the day said, "shout." They were not out to awe the tired businessman and his wife with their artistry; they merely wanted to create a pleasant rapport and possibly send ticket buyers home with a little ditty that could be carelessly hummed. Finding that little ditty, especially if the performer could not write his own, was more important than creating a well-thought-out, integrated score. As a result, stars and producers, who had by this time supplanted the fast-retreating comic opera ensembles, were willing to incorporate any composer's song if it promised them success.

The best examples of this kind of interpolation can be seen in the early, practically parallel careers of Blanche Ring and Marie Cahill, both of whom were catapulted to fame by interpolations and both of whom thereafter actively sought such special material. Miss Ring's rousing rendition of George Evans and Ren Shields's "In The Good Old Summertime" made Charles Denee's official score for *The Defender* seem pallid. Later she had similar success when she interpolated "Bedelia" and "I've Got Rings On My Fingers" into other men's scores. Ludwig Englander's score took a back seat when Miss Cahill belted out Clifton Crawford's "Nancy Brown" in *The Wild Rose*. Four months later she diminished another Englander score when she introduced Bob Cole's "Under The Bamboo Tree" in *Sally in Our Alley*.

Because of such examples, interpolations, which had been managed with some restraint in imported comic operas and which usually had been prohibited in better domestic ones, now became something of a plague, albeit one with a far-reaching precedent in the piecemeal musical comedies of the past. By the time Julian Edwards's *The Girl and the Wizard* reached Broadway it included only six of his original songs and had eight interpolations, representing the work of five composers. Inevitably, certain stylistic clashes ensued. Coupled with the pasted-together quality of all but the best musical comedy librettos, this told against the period's entertainments. Much the same might be said of the era's revues, although something of a theatrical kaleidoscope was to be expected in them.

Of course, operetta had not totally disappeared from the footlights. In America, many of the older composers and librettists remained active. Indeed, Herbert produced one of his best works, *Mlle. Modiste.* Though not precisely typical of Herbert's shows, at least qualitatively, it had one of the few superior librettos Herbert ever wrote for, Henry Blossom's engaging tale of a hat shop salesgirl who becomes a great opera star and marries the handsome officer she has long loved. It had, as well, the captivating Fritzi Scheff to sing one of Herbert's finest melodies "Kiss Me Again." Other great Herbert melodies from the show, which were awarded to other performers, as was his custom, included "I Want What I Want When I Want It," still a popular baritone and bass solo, and "The Time And The Place And The Girl."

One new figure, Gustav Luders, had appeared on the scene. Luders, who generally collaborated with his favorite lyricist and librettist, Frank Pixley, was something of an oddity. He came in at the end of comic opera's first heyday and thrived briefly before operetta's second flourishing. His was a minor talent, one that never seemed to enlarge (although De Koven, and possibly Edwards and Sousa, might also be charged with a lack of artistic development). Yet in his own restricted way Luders wrote beautifully and was tremendously popular. One quirky evidence of his popularity remains even today, for comic opera lovers, seeking out old vocal scores in secondhand bookshops and other byroads, almost certainly come across more copies of his masterpiece, *The Prince of Pilsen*, than of any other work of the epoch.

In one respect Luders represented a bridge between the two periods, for, unlike any of his important predecessors, he was born in Germany. (His contemporaries, Ludwig Englander and Gustave Kerker, were also born there, but they generally composed in a slighter vein, decorating the musical comedies and revues of their day.) Luders's music was lighter as well, less florid than other comic opera composers' work, but nevertheless clearly in operetta and not musical comedy tradition. In this respect, too, he served as a bridge. Of course, Luders offered his share of thumping marches, sweeping waltzes, and sweet lullabies, as well as all other forms dear to his audiences. Good German that he was, he savored the music of the old beer halls. His most famous song was probably *The Prince of Pilsen*'s "Heidelberg," or as it became popularly known, "The Stein Song." It was a forceful enough piece in a tradition that took in *Robin Hood*'s "Brown October Ale" and, later, *The Student Prince*'s "Drinking Song."

The period's critics often employed "dainty" as one of their most commendatory adjectives, and "dainty" is precisely the word that best describes Luders's most characteristic writing. From his first successful works, *The Burgomaster* and *King Dodo*, to his last, most Luders scores included at least one "tale" or "message" song. *The Burgomaster*'s hit had been "The Tale Of The Kangaroo"; *King Dodo* included "The Tale Of The Bumble Bee." *The Prince of Pilsen* offered two: "The Message Of The Violet" and "The Tale Of The Sea Shell." Their forms varied. For instance, "The Tale Of The Kangaroo" had a verse in 4/4 time, a chorus in 2/4; "The Tale Of The Sea Shell" was entirely in 4/4; the most famous of all, "The Message Of The Violet," was a waltz. Yet however variegated their formal structure, their essence was one of quietly reserved delicacy. That they possessed a certain daintiness is incontestable. The roots of "dainty" go back to the same Latin word that gave us "dignity," and these songs also have a charming, understated dignity.

Yet if musically Luders cast a soft light on things to come, the librettos for which he wrote remained glaringly traditional. Pixley, as well as Luders's other associates, rarely moved back to bygone times, but their settings were often sufficiently far away and, when they were not, the stories were sufficiently farfetched to effect the romantic mood comic opera by then demanded. *The Prince of Pilsen* was set in contemporary (1903) Nice. A vacationing Cincinnati brewer is mistaken for the prince of Pilsen, a confusion the prince allows to continue since it offers him an opportunity to court the brewer's daughter without disclosing his identity. *The Burgomaster* was set mainly in 1900 New York, but its story told how Peter Stuyvesant and his secretary, Doodle Van Kull, return to the city to gape at their modern descendants. In this insistence on traditional comic opera fare, Luders was unwittingly falling into the very pattern American operetta writers would continue to pursue—to pursue, one must add, in the face of innovations from overseas. Less than a month after *The Burgomaster*'s New York premiere on December 31, 1900, Luders and his American competitors might have glimpsed the wave of the future.

On January 23, 1901, Rudolph Aronson brought Johann Strauss's *Wiener Blut* to Broadway under the title *Vienna Life*. By 1901 Aronson had long since lost his Casino, and his career as a producer had become erratic. His motives for producing this work, which Strauss had not lived to complete, are obscure. No doubt he saw an opportunity to profit from some momentary nostalgia evoked by the composer's death. Moreover, he could, for the first time in years, personally offer American playgoers an operetta whose world premiere had delighted Europeans the preceding season. But in view of his emotional, sentimental, and insistently optimistic nature he may have hoped he also could revive the palmier times he had once enjoyed.

Whatever his reasoning, Aronson's decision was courageous, and foolhardy. Americans had had their surfeit of Viennese operetta in the eighties. In the eight theatrical seasons before Aronson mounted *Vienna Life*, only three Viennese or German musicals had come to New York. Two—*The Queen of Brilliants* and *A Dangerous Maid*—had been so thoroughly rewritten, so crammed with American business, and so retextured with domestic musical interpolations that the authors of their Middle-European originals would have had difficulty in recognizing them. The third operetta, *Lilli Tsi*, while apparently mounted with reasonable fidelity, was only a one-act curtain raiser. All three were failures. Not since the 1891–92 season when Carl Zeller's *The Tyrolean* (*Der Vogelhandler*) and Carl Millöcker's *The Vice Admiral* (*Der Vice-Admiral*) were counted among Broadway's hits had New York theatregoers enthusiastically flocked to see Viennese operettas.

So Aronson all too quickly learned that playgoers were not as moved by roseate memories as he. In his autobiography he recalled ruefully, "Although the production was magnificent and the cast included Raymond Hitchcock, Ethel Jackson and Amelia Stone—the theatre-going public kept steadly away, having been inoculated with musical comedy and rag-time and preferring that class of entertainment to the old-time Viennese Operetta." Aronson withdrew his presentation after thirty-two poorly attended showings. A few months later *Vienna Life* appeared for a week as part of the ever changing bill at Henry Savage's Castle Square Light Opera Company. Whether Savage had secured Aronson's discarded sets and costumes, and whither the mounting moved after its eight-performance revival, if it moved anywhere, cannot be readily determined.

Had *Vienna Life* proved a success, Aronson might have been hard put to find a follow-up, for by 1900, death had claimed not only Strauss but virtually all the other composers of the early Viennese and German schools. Many of the librettists were dead or retired, too. Significantly, *Wiener Blut*'s librettists were younger men, whose conceptions of plotting were not irretrievably bound to the past. As a result, Viktor Leon and Leo Stein rejected a Graustarkian setting and a tale of sweet princely romance, electing instead to place the action of their modern bedroom farce in turn-of-the-century Vienna. The fickle Count Zedlau is representing his country at the Vienna Congress. He is separated from his wife, lives with a dancer, and makes advances to his servant girl. The count's prime minister somehow mistakes the countess for the dancer and begins a courtship of his own. In the end, of course, everyone is properly paired.

To the extent that so many earlier Viennese librettos had unfolded what were bedroom farces at heart, Leon and Stein's choice was not totally original. They even incorporated a quiet bow to Graustark by allowing their principals to present the Ruritanian land of Reuss-Scheiz-Greiz. But they set all their action in a recognizably contemporary Vienna. This, too, was not new. Several major Viennese operettas had done this before, notably *Die Fledermaus*. In this instance, however, Leon and Stein were, knowingly or not, establishing a precedent which they were to consolidate five years later when they collaborated with Franz Lehar on *Die lustige Witwe*.

5

The Merry Widow and the Viennese Waltz-Operetta

"To-morrow night is coming to town a young person who has attracted an enormous amount of attention in foreign parts and who is expected to attract as much here. Whether she does or not, however, great preparations have been made for her, and her wiles are expected to be potent. She is 'The Merry Widow' who is the heroine of Franz Lehar's comic opera of that title." So began the extended welcome the *New York Times* proffered on October 20, 1907, the eve of *The Merry Widow*'s Manhattan premiere. That welcome went on to include not merely a history of the show and its composer, as well as of the principals of the new mounting, but a detailed synopsis of the plot and lengthy musical excerpts to boot. It was not the sort of welcome the *Times* extended very often. Clearly something was up.

The Merry Widow could not have waltzed across the world's stages at a more propitious time. It came as close to being the perfect turn-of-the-century stage piece as anyone could have hoped. Admittedly it was slightly flawed: its lyrics and dialogue were not of the very highest order. Even in the original German, they lacked the biting wit and stylish literacy of almost anything Gilbert had put his hand to. And no foreign adaptation was able to transmute their base metal into gold. But this flaw is more evident in retrospect than it was to critics and audiences of the day. Still, the words, in any language, were good enough. If they hardly helped, they didn't hinder. What mattered was the story, the attitudes underlying that story, and, most of all, Lehar's meltingly beautiful melodies. More than anything else, the music carried the day—as it still does in any revival—and almost assuredly would have triumphed attached to any tale. The libretto, accompanied by a lesser score, also might have succeeded, though obviously not as rapturously. One has only to remember typical scores of the time to realize that success did not always derive from greatness.

The totality was irresistible. In fact, the rather ordinariness of the lyrics and dialogue may have added to that totality. Gilbert's brilliance was often untranslatable, even when it was not singularly Anglo-American in taste. But Leon and Stein's more commonplace sentiments could be made at home anywhere in the Western world. And in small time they were. No careful and precise records exist of the number of performances *The Merry Widow* enjoyed, especially in its early years, nor of its grosses. But it is safe to say that the show has been played more often than any other musical ever written. Within a few seasons of its Vienna premiere the piece was being performed all across Europe—in St. Petersburg, Trieste, Milan, Oslo, and Bucharest, to cite a few examples. The United States, South America, Australia, and the more Occidental enclaves of Africa and Asia all avidly welcomed it. As French musical theatre historian Florian Bruyas noted, without a trace of French immodesty, "It remained only for Paris to bestow its blessing." That blessing was granted in April 1909, and, as Bruyas continued, "A revolution convulsed operetta."

Operetta was not alone in changing. The world into which *The Merry Widow* made so fetching an entrance was a world in flux—and enjoying every minute of it. The Progressive Era, the Edwardian Era, La Belle Epoch—call it what you will—was a period of giddily expanding horizons. Occasional setbacks, sporadic displays of human ugliness could be waved away with a genuine absence of cynicism, a heartfelt certainty of moral and social rectitude. These feelings pervaded virtually all of the Western world, but nowhere more than in the United States, and there nowhere more than among the middle and upper-class New Yorkers who comprised *The Merry Widow*'s audiences.

The very theatre in which these audiences sat exemplified this plush, secure new world. The ornate, pseudo-baroque New Amsterdam, flagship of the Erlanger empire, was less than four years old when *The Merry Widow* opened there. It was the fourth house built on 42nd Street, the first having been the Victoria in 1899. Not only did the New Amsterdam consolidate the importance of 42nd Street as a principal theatrical thoroughfare, but, situated as it was a few steps from Seventh Avenue, it helped make Longacre Square, later Times Square, the heart of a new theatrical district. Forty-second Street had also acquired electricity a mere dozen years before, and just the previous season Charles Dillingham had given Broadway its first advertising sign with moving illumination.

If many in the audience came to the theatre from their brownstone or brick row houses, a steadily increasing number came from spanking new apartment houses, which were rapidly supplanting nineteenth-century boardinghouses. A few of the "gallery gods," having scrimped to purchase their cheap seats, may have come from tenements recently protected by the 1901 Tenement House Law, a law that soon served as a model across the nation. While tenements were not to disappear from New York, the law at least offered hope that tenements would be provided with better light, ventilation, fire protection, and sanitation.

A handful of playgoers may have taken advantage of nice weather to walk from their residences to the theatre, but most probably came by horse and carriage, horse-drawn buses and trams, or the smoky, clattering elevated lines. In 1907, no small number undoubtedly availed themselves of the three-year-old subway, which took mass transportation underground and thereby allowed another newfangled convenience, the automobile, room to assert a seemingly irrevocable ascendency.

Away from the theatre, evidences of twentieth-century drive and achievement were everywhere. Theatres were not the only buildings rising. The Flatiron Building, constructed in 1900 as the Fuller Building, and the even taller Singer Building, under construction in 1907, gave instant and wide currency to the term "skyscraper." In New York, these two and other high-rise office complexes housed corporate headquarters for seventy per cent of the nation's largest companies. Prospects for all these industries seemed bullish, while regulatory agencies, which feisty Teddy Roosevelt was ramming into existence, promised to see that corporate growth would coincide with a new sense of corporate responsibility.

Although "streamlining" was a term several decades in the future, the arts were quietly doing just that. Victorian gingerbread was beginning to appear excessive. The more fluid, cleaner "art nouveau" was the rage, while still simpler, relatively more austere works of art, of a style later known as "art deco," were on the drawing boards and in the ateliers of the avant garde. New movements in American painting were typified by a group of young painters who had recently moved from Philadelphia to New York. Their subject was the everyday life about them; their brushwork was brash and free. In time they would be called the "ashcan school." Prose, both in literary works and in the era's periodicals, still tended to be verbose and florid, but a writer such as Stephen Crane already had offered an example for the future.

The ethnic composition of New York continued to change, and the number of blacks had dropped a bit. Large influxes of Irish, Germans, and Jews, who had arrived in the nineteenth century, began to consolidate their power and move into the mainstream. These three groups were avid theatregoers. Together they were moving into positions of importance in the theatre and altering its very texture. Playgoers of English stock might welcome *The Merry Widow* intellectually as a fine piece of popular theatre, but the Germans and Jews would bring additional emotional ties to its art, while the Irish would embrace anything capable of pushing English works out of the limelight.

Inevitably, there were those who did not see this change as progress. Voices lamenting the passing of old styles, old ways, old social structures cried out at intervals. Newspapers and books of the time duly recorded their wailings and, more often than not, moved on to ignore them. The forces of movement were not to be thwarted. Yet it would prove a curious virtue of *The Merry Widow* that it offered both the modernists and the traditionalists something to cheer about, although this was not always clearly perceived at first.

Those with a sense of theatrical history could see the piece as the latest in a long honorable line. The *Times*, continuing the article quoted earlier, recorded the operetta's European acclaim, an acclaim described as "unprecented for a good many years—since the times, in fact, that wrinkled, gray-headed men, in mentioning comic opera, speak of with a sigh as 'golden.' " The article, becoming in turn facetious, wistful, and finally hopeful, added:

> They mean sometimes as much as twenty years ago. The very aged can remember almost thirty years back, when there was a piece called "Pinafore," and earlier and later they can remember such other things as "The Mikado" and "La Fille de Madame Angot" and "La Belle Helene" and "Die Fledermaus" and "The Black Hussar" and "Fatinitza" and "The Mascotte," and others which they could mention if their memories weren't going so fast. There were comic operas in those days, and they had an immense popularity, which they mostly deserved. But there has been a long and weary drought since the "golden days," in which "musical comedies" and other such things imported from England and produced at home by our distinguished "Broadway school" of composers have prevailed, which have by no means quenched the thirst of the aged and which may well have left a vague, unsatisfied longing in the mouths of the young.
>
> Now if "The Merry Widow" shall prove to be a real heir of the ages and bring back something more than memories of the golden days, there will be rejoicing within a long radius of Broadway.

Two days later the *Times*'s critic assured his readers that the show was indeed a legitimate heir when he proclaimed that its music "carries it into a class which compels one to remember 'Fledermaus' and 'La Grande Duchesse' to find anything to compare it with."

Theatre Magazine's critic, not quite as taken, nonetheless acknowledged a similar pedigree, relating this latest offering to older and newer works while adding, gratuitously, a rather snobbish aside. He observed, "It is not nearly so beautifully made a score as Messager's 'Veronique'—but this latter work was not for the large masses—nor is it filled to overflowing with melody as 'Mlle. Modiste.' " Such reservations aside, he granted that *The Merry Widow* was "one of the most charming works of its caliber seen here in many a musical moon."

Less explicitly but with more satisfaction, the *Herald* announced that Lehar "has found the long lost trail of melody. He has followed it to its lair and has given the world some fine, tuneful game." (Its arrival coincided, after all, with the start of the hunting season.) Along with most of his fellow critics, the *Herald*'s man also had complimentary things to say about the libretto:

> And it was all about something! There was that elusive thing, a plot, present. The music illustrated something vital and human, for the attention of the audience followed the unveiling of the story as it seldom does. . . . A lot of conventional situations were employed, but the outcome was excessively interesting.

Modernists, while not denying the magnificence of the music, pounced on the libretto as an exemplar, hailing it for pruning away antiquated folderol. To a large extent, New York would continue to see Viennese operettas as superior, usually blaming whatever faults were discerned to careless or tasteless American translators. For instance, five years later, *Theatre Magazine* opened its review of *The Rose Maid* by noting, "There is always a story of some consistency and sanity in the librettos written for foreign operas." The *Dramatic Mirror*, the era's leading theatrical trade paper, celebrated *The Merry Widow*, especially its libretto, as a major achievement:

> Coming at the end of an epoch of inane musical comedy—grant that it is at an end!—the operetta is twice welcome, on account of its own excellence, and because it may start a new era in musical entertainment. The music is bright and original, the humor fresh and genuine; the story clear and vigorous and the characters exaggerated only a trifle beyond probabilities.

Echoing those sentiments, the *Times*'s critic embraced *The Merry Widow* as "the greatest kind of relief from the American musical comedy."

The *Dramatic Mirror* was often short-sighted, but in suggesting that *The Merry Widow* was opening a new era in musical theatre the paper was right on the button. For the next seven years Viennese-style waltz-operetta dominated the American lyric stage. This domination appears obvious in retrospect partly because the great operettas of the period are almost its only theatrical survivors and partly because a number of theatrical historians have firmly convinced later generations of it. In his autobiography, Harry B. Smith wrote, "Lehar's opera made a fortune for Colonel Savage, its producer. Eventually it cost American managers [as producers were then called] a number of fortunes, because for several years afterward, they produced every German and Austrian operetta they could get, in the vain hope of finding another 'Merry Widow.' " Later in his reminiscences Smith went so far as to suggest Austrians and Germans wrote operettas "mainly with the idea of selling the American rights." This suggestion is plainly preposterous, for European success was rewarding enough in every way. But Smith vividly exemplifies the impression Viennese operetta and its success made on observers of the day.

Yet statistics, coupled with the attitude of many of the era's theatrical figures, imply the dominance was not quite as real or as fully understood as time suggests. Take the statistics. Numbers alone would appear to dispute Viennese operetta's supremacy. Of the nearly one thousand plays presented on Broadway between mid-June 1907 and mid-June 1914, just over two hundred and fifty were musicals, while only some thirty of these could be considered Viennese or German operetta. Three per cent (twelve per cent of musicals) hardly constitutes an incontestable proof of dominance. But their record of success tells a significantly different story. Only one-third were outright failures—a remarkable average by theatrical lights. A little over a second third were borderline cases, in several instances failing to please New York but recouping their costs and their reputations on the road. The remainder—nearly a third of the importations—were among the smash hits of the era.

The long-run musicals of this epoch (excluding the *sui generis* extravaganzas mounted by the Hippodrome) were *The Merry Widow* with 416 performances (which accomplished the remarkable feat of outdistancing even the Hippodrome's seasonal entry), *The Pink Lady* with 312, *The Chocolate Soldier* with 296, *The Dollar Princess* and *Three Twins*, with 228 performances each. Oscar Straus's *The Chocolate Soldier* and Leo Fall's *The Dollar Princess*, along with Lehar's work were, of course, Viennese. *The Pink Lady* and *Three Twins* were American, although the composers of both were European born and trained. In sound and sentiment *The Pink Lady* was a practically perfect imitation of Viennese models. *Three Twins*, however, never ventured far from the period's accepted musical comedy styles, even if one can occasionally catch Middle-European inflections in Karl Hoschna's score. In short, while only twelve per cent of the musicals produced during these seven seasons were Middle-European importations, they and their American imitations accounted for four of the five biggest successes. Several others placed high on the list of runners-up.

If statistics, then, slightly confuse the issue of dominance, many of the day's theatrical figures further muddied the waters. Their treatment of the shows implied that distinctions between musical genres were either misunderstood or purposely fudged. Although the *Times*'s banner called *The Merry Widow* a comic opera and its pre-opening article related the new work to the older tradition, both the article and *Times* review also employed the term "operetta" as a synonym. One can distinguish between earlier comic opera and later operetta, and, though at this point the distinctions were not that apparent, the juggling of both terms was not necessarily indiscriminate.

Difficulties in definition of terms began to creep in three months later when Oscar Straus's *A Waltz Dream* premiered. Perhaps even more than *The Merry Widow*, *A Waltz Dream*'s heart and soul were deeply embedded in Middle Europe. Straus composed an exquisite score for the story of a young man's unhappy marriage to a foreign princess and of his sneaking off to dream to the music of his homeland, played by an all-female orchestra in a park near the palace. "Love's Rondelay" ("Leise ganz leise") was merely the most instantly approachable of a melody-packed musical program.

Like *The Merry Widow*, *A Waltz Dream* was advertised as an operetta and treated as such by most commentators. Even the *Times* was almost always consistent in so calling it. Almost but not entirely. In summing up his satisfaction with what was obviously about to become a trend, the paper's critic observed, "No one need be sorry that the tendency nowadays is to turn to Vienna for musical comedy." If, for the sake of tradition, conservatives might choose to call *A Waltz Dream* a comic opera, as they had *The Merry Widow*, there would have been little to fault. But one thing *A Waltz Dream* was not was musical comedy! One would like to think that the *Times*'s man let the expression slip by in a moment of carelessness, or that he utilized it as part of a sweeping generality that would not be accepted literally. But most likely he used the term deliberately, sharing the conceptions of his contemporaries. The chicness of the Gaiety musical comedies that ushered in the new century had stamped them indelibly as the ultimate form for a bright new day. They had left the older comic opera seeming musty with age. To turn the tables on musical comedy so soon and to declare it obsolescent was perhaps asking too much.

Within a season or two, nearly half the operettas that appeared on Broadway called themselves musical comedy or, failing that, sought for some novel term that would not tie them to the older terms, comic opera or operetta. *The Dollar Princess* is illustrative. By the standards of the day the operetta was as modern as all get-out. Not only did it employ a contemporary setting, but that setting was the American business world. The story also dealt with a number of romances, the most important ones being the infatuation of an American business tycoon with a supposed countess, who is actually a nightclub singer, and the tempestuous love affair between the tycoon's daughter and a young business executive. The daughter believes she can buy any man she chooses, a philosophy which alienates the young businessman. Money moves all the action of the plot. The tycoon must pay dearly to rid himself of the phony countess, and the young executive, to show he still loves the tycoon's daughter, offers her father his thriving business for a nominal price.

When Charles Frohman brought *The Dollar Princess* to Broadway in 1909 its billboards, its programs, its sheet music all promised theatregoers the very latest and best in musical comedy. Of course, Frohman had been presenting Broadway with the latest and best in musical comedy since 1895, and no doubt he hoped to capitalize on a sense of continuity. When his partners, Marc Klaw and Abe Erlanger, presented *The Pink Lady* they followed his lead, calling it too a musical comedy. *Madame Sherry* was offered to playgoers as "a musical vaudeville." Moving in a totally different direction, perhaps as a bow to his star Emma Trentini, Arthur Hammerstein listed Rudolf Friml's *The Firefly* as "a comedy opera." But *The Firefly* was American. Few

Viennese importations followed its lead or that of *The Chocolate Soldier*, which termed itself a comic opera. Several operettas were billed simply as musical plays.

An old unwritten rule of the theatre, mentioned earlier, was that comic opera scores, especially those by the recognized masters, were granted a certain sanctity and thus escaped the prevailing plague of interpolations that were commonplace in musical comedy. Omnipresent interpolations in the operetta scores of this period therefore hint, among other things, at the perception of these scores as musical comedy. Even Lehar's works were not free from tampering. Luckily, his greatest scores—*The Merry Widow, The Count of Luxembourg,* and *Gypsy Love*—were heard as pure Lehar, but his weaker scores for *Eva, The Man with Three Wives,* and *The Maids of Athens* were supplemented by American "improvements." Apparently, no important American composer was asked or consented to make the additions, so the chance of any of the interpolations saving the day for these lesser shows was meager. And none did. Who now remembers Al Brown's "Tale Of The Jealous Cat" or Paul Kerr's "Our Glorious Stars And Stripes"?

Leo Fall's five imported operettas were second only to Lehar's six. Americans rightly felt Fall lacked Lehar's melodic resources. As a result, Fall's scores were amended from the start. Even his masterwork, *The Dollar Princess,* came to New York with American novelties added. Most of Fall's operettas were brought over by Frohman, who, happily for Fall, had employed a great composer to devise many of the additions. Jerome Kern was not yet famous, of course, nor had he fully developed his supreme creative gifts, but his Bohemian and German heritage and his impeccable taste saw to it that his interpolations clashed only slightly with Fall's musical mannerisms. To the extent that Kern's melodies did clash, they drew the music away from operetta toward the better musical comedy writing of the period. And for all the disparities of style, Kern's sensitive contributions were a far cry from, say, the incontestably American "coon" song, Chris Smith's "I Want A Little Lovin' Sometime," which Marie Cahill so jarringly inserted into Richard Heuberger's exquisite *Opera Ball.* Such material, especially when it contained a pronounced ragtime lilt, moved the music still farther along the road toward musical comedy writing.

In several ways *The Merry Widow* and its American mounting may have unwittingly promoted the confusion of genres right at the new era's very start. First, its story was not that different from the tales unfolded in the Gaiety musical comedies or their American imitations. Apart from an unimportant subplot in which the Pontevedrinian (Marsovian, in English translations) ambassador's wife has a brief flirtation with a French officer—allowing him to sing "Sieh dort den kleinen Pavillon"—the story told how Count Danilo, secretary to the Pontevedrinian embassy, is pushed to woo and wed the rich widow Hanna Glawari (Sonya, in English translations) so that her fortune will remain in Pontevedro. Danilo and Hanna had once been in love, but Danilo's uncle had threatened to disinherit him, so Hanna had wed a banker. Danilo has not married, preferring to pass his time with the grisettes at Maxim's. At least that is what he insists on in public as his preference. He soon realizes, however, that he is still in love with Hanna, and she with him.

The Merry Widow's plot demonstrated that a continuing social leveling was under way. Deities and royalty were less and less the concern of operetta librettists, while musical comedy storytellers, at least in the "classier" entries, had abandoned lower-class roughnecks and gauche arrivistes. Kings and queens gave way to princes and princesses, generals to lieutenants. Princes might still woo and wed shop girls, and young officers court laundresses or lady fiddlers, but when they did not in the end reject them for one of their own set, the story made it clear that the shop girl or laundress was either a princess in disguise or a youngster of such inherently noble character that she deserved elevation.

In outline, then, there frequently was little that separated one genre's librettos from the other's. What did differentiate the two was the tone set by the librettist. Writers of operetta sought to convey a sense of high romance and a certain stagy earnestness; musical comedy became far more determined to be comic than was comic opera. Not that operettas were devoid of comedy. Their frequent recourse to Middle-European settings and characters offered dialect comedians a field day. In both musical comedy and operetta, romance, love, and laughter unfolded for the most part in fashionable, contemporary cities—real, recognizable watering holes. Paris, Vienna, even New York frequently replaced Graustark and Ruritania.

Even musically *The Merry Widow* edged toward lighter entertainment. Magnificent as Lehar's melodies and orchestrations were, they were a step or two down from earlier comic opera's grand opera antecedents. Lehar's musical lines were softer, less given to the startling leaps or fioritura passages so common in late-nineteenth-century works. Listening to Offenbach or Strauss or Sousa with one's eyes closed, one sometimes might have imagined oneself in an opera house; listening to Lehar, one knew one was in a legitimate theatre or on a dance floor. For *The Merry Widow* changed dancing both on the stage and off. Set aside more or less forever were the regimented drills and classic ballets that had long dominated our lyric stage. From *The Merry Widow* on, softly swaying waltzes became the order of the night. The principal love song and hoped-for hit of every new operetta was a ballroom waltz. Not unjustly, many historians see in these incomparable waltzes the seeds of what soon evolved into a rage for ballroom dancing, into what was called "the dancing craze."

Though no other song quite matched the vogue and influence of *The Merry Widow*'s great waltz, the rest of Lehar's score quickly attained a much merited popularity. Hanna's "Pontevedrinian folk song," "Vilja," had a haunting, languid grace. Danilo was assigned two rousers, "Da geh'ich zu Maxim" (generally called simply "Maxim's" in English) and "Ja, das Studium der Weiber ist schwer" (usually "Women" in English). In their own small ways, these songs, too, probably helped lighten the texture of comic opera music.

These changes undoubtedly prompted a third change, which may have been the most significant of all in the minds of *The Merry Widow*'s first audiences. A basic change in the nature of casting occurred, which may have originated not in New York but in London. For the part of Danilo, Savage turned to a rising song-and-dance man who had appeared with George M. Cohan and who was more often applauded for his fast stepping than for his singing. "Light of voice and lighter of feet," the *Herald* would characterize him. Supple, boyishly handsome Donald Brian was a new breed of operetta hero. Although he initially sported a mustache to give him a certain dignity and maturity, he looked ridiculous and hurriedly dropped it. He was, in any case, a far cry from the hulky, lumbering leading men of a fast receding school.

But Brian was not the very first such hero. A suspicion remains that Savage may have taken his cue in casting Brian from George Edwardes, who had awarded the part to another American song-and-dance artist, Joe Coyne, for *The Merry Widow*'s London production. Coyne, lacking even Brian's passable voice, simply recited his lyrics. Legend insists that Edwardes persuaded Lehar to attend the London rehearsals, told the composer his leading man had a sore throat, and at run-throughs noisily begged Coyne to save his voice by talking the songs. The truth supposedly came out at the dress rehearsal, leaving the producer to calm the outraged composer. Edwardes is said to have argued that Coyne was a very funny man, which led Lehar to retort that he did not write funny music. Edwardes's biographer, Alan Hyman, records that Lehar changed his mind about Coyne's interpretation after seeing the opening night audience's response and thereafter suggested that all Danilos recite their lyrics. Be that as it may, Coyne and Brian established that a leading man need not have an operatic voice to sing in operetta.

Nor was the singing of Ethel Jackson, the female lead, of a high order compared to earlier prima donnas. Critics clearly preferred her acting to her singing. The *Times*, for example, dwelt on the superiority of her vivacious interpretation to the demure performance of London's Lily Elsie, adding, "She makes the waltz the dramatic moment in the action as it should be." Thus while Coyne and Brian were altering the image of operetta heroes, Miss Jackson was doing much the same for operetta heroines.

Not every imported operetta rushed to hire this new style of principal. A *Waltz Dream*'s Nicki, the first Viennese lover to follow Prince Danilo to our shores, was initially sung by a fine singer with grand opera ambitions. Indeed, the original American Nicki, Edward Johnson, went on not merely to sing in grand opera but to culminate his career as manager of the Metropolitan Opera. Nor was he totally unique in his excellence. Now all but forgotten names such as Ann Swinburne, Ida Brooke Hunt, and William Pruette brought their superb, well-trained voices to enhance a number of lovely Viennese scores. Nevertheless, a pattern had been set. Time and again critics were forced to emphasize a performer's charm or stage presence and could only fall back on a few discreet adjectives to describe voices they deemed "dainty," "pleasant," "spirited," or, less kindly, "suitable" and "adequate."

So, apart from some early comments relating *The Merry Widow* to the preceding generation's beloved comic operas, the operetta was welcomed for the excitement of its novelties as much as for its inherent excellence. What appears to have gone unperceived at the time was *The Merry Widow*'s curious dualism. All the while it flashed its modernity—its chic figures, played by its new style of performers, its contemporary settings, its plot pruned of antiquated excesses, its softened musical lines—all that while it quietly exuded wisps of nostalgia—a preoccupation with a decorative nobility, passing bows to a dying world of small Ruritanian principalities, a gentle waltzing in an age moving to harsher and more rapid tempos.

A financial panic struck Wall Street two days after *The Merry Widow* opened, temporarily clouding some people's sunny futures. The glittering success of the new musical in the face of so brutal a fiscal crunch did not pass unnoticed. Its attractions as art and escapism clearly overrode all but the most vexatious budgetary demands. Happily, the panic passed with reasonable dispatch. Good times, times for new and better things, returned. But Broadway thought it had been taught a small lesson. Waltz-operetta meant money in the box office.

6

American Operetta's Response to Vienna

Americans did not dash headlong to mimic Viennese styles. In fact, for the seven seasons following the premiere of *The Merry Widow*, Americans wrote far more musical comedies and revues than they did works that could be called waltz-operettas. Moreover, a few Americans had never really stopped writing earlier-style comic operas. There had always remained some sense of continuity with the older traditions. In 1905 Victor Herbert's *Mlle. Modiste* was so effusively welcomed that many thought Herbert would never compose a superior work. Herbert never really abandoned the musical stage, except to head the Pittsburgh Symphony for three years. De Koven kept plodding along, never with more than middling success. Luders joined the lists. Edwards contributed one or two final, if feeble, efforts. Ivan Caryll and Karl Hoschna were to straddle the fence between musical comedy and operetta. At almost the end of these seven seasons Rudolf Friml appeared on the scene. Sigmund Romberg came to Broadway after Friml, and his gifts as an operetta composer did not come to light until much later.

In the 1907–08 season De Koven's *The Girls of Holland* and

Edwards's *The Gay Musician* displayed the flag for native operetta. Both were quick flops. (An English comic opera, Edward German's *Tom Jones*, had only slightly better luck.) The season following *The Merry Widow*, only one Viennese operetta, *Mlle. Mischief*, was brought over. It enjoyed a modest reception. Competing with *Mlle. Mischief* and with twenty-eight American or English musical comedies and revues were a mere four domestic operettas—Luders's *Marcelle*, De Koven's *Golden Butterfly*, and two by Herbert, *Algeria* and *The Prima Donna*. A third Herbert opus of the time, *Little Nemo*, exemplified the undeterminable nature of many of the period's lyric pieces. The work was billed as "a musical comedy extravaganza." Its musical range was far smaller and its style lighter than Herbert's material for either *Algeria* or *The Prima Donna*. Herbert himself spoke of it as music "of a popular order." In his brilliant analysis of the work, Herbert's biographer, Edwin Waters, calls it both musical comedy and operetta.

Not until the last three seasons of this epoch—1911–12, 1912–13, and 1913–14—when producers seemingly imported Viennese operettas wholesale, did the number of comic operas begin to approach or exceed the number of musical comedies and revues. Yet even in these seasons the number of American operettas never passed seven or eight. Without Herbert's contributions the total would have been appreciably lower and a sense of lasting value significantly less.

Set side by side with the Viennese school of operetta, the American works disclose some strikingly independent, and perhaps surprising, characteristics. When watching or hearing them, one often has an inescapable sense of *déjà vu* or *déjà entendu*. They seem resolutely aligned with the older order of comic opera, rather than with the relatively more worldly, chic, and gay atmosphere of the Viennese school. The callous, mercenary love of *The Merry Widow*'s grisettes, the money-grubbing machinations of *The Dollar Princess*'s modern business world, the scruffy Parisian Bohemia of *The Count of Luxembourg* are, more often than not, lacking. Not that all these things were not romanticized to some extent in these Viennese confections. But clothed as they were in contemporary dress, they brought matters home a bit more. They were never underscored; they were there for audiences to catch or ignore as they chose, and no doubt many chose to ignore them. Villainy, greed, and all the other real and theatrical evils were the stock-in-trade of American librettists, too. But American writers frequently garbed the villainy in such heightened romantic colors that it seemed totally theatrical, and thus totally undisturbing outside the context of the evening. Wives might worry about husbands trafficking in the demimonde or assuming the risks of high finance, well-heeled men and women alike might shrug sadly at the tacky poverty of some artists, but they would never take seriously the wiles of *Naughty Marietta*'s black-mailing pirate or the treachery of the unctuous baron in De Koven's *The Golden Butterfly*. A considerably larger proportion of American librettos were set in romantically far-off times or colorfully exotic places. Viennese operetta approached this same picturesque ambience largely in its treatment of gypsy motifs.

One reason so many American librettos may have taken the tack they did was that the leading librettists were carry-overs from an earlier theatre, like several of the leading composers. That Harry B. Smith's hand was obvious in a hefty number of these librettos should surprise no one familiar with the history of our musical stage. Smith's friend and rival, Stanislaus Stange, also contributed a notable share. Tellingly, when plots did give off a more modern air, they were usually the work of a younger generation of writers, of which Otto Harbach and Henry Blossom were probably the best.

Harbach and Blossom came from disparate backgrounds. Blossom was the older of the two, having been born in St. Louis in 1866. Harbach, whose original name was Hauerbach, was born in Salt Lake City in 1873. Blossom came from a monied family and attended private schools, but spurned college to enter his father's insurance business. Harbach's parents were Danish immigrants who sent him to Knox College, in Galesburg, Illinois, where he prepared for a teaching career. Boredom prompted Blossom to abandon insurance for writing; failing eyesight forced a similar change on Harbach.

Henry Blossom gave Herbert the best librettos he was to have. Herbert must have been partially aware of this, for while he used Smith as a librettist in eighteen of his shows and Blossom in only eight, he apparently moved quietly away from Smith once he was satisfied with Blossom's abilities. Smith, then, wrote virtually all of Herbert's early librettos, but gradually did fewer and fewer until their next-to-last collaboration on *The Enchantress* in 1911. Blossom continued to supply Herbert with librettos until Blossom's premature death in 1919. Actually, apart from the 1919 musical comedy *Angel Face*, after Blossom's death, Smith's collaboration with Herbert effectively ended in 1908. His work for *The Enchantress* and *The Duchess* was largely on their lyrics, with only minor contributions in rewriting other people's librettos. *Angel Face* aside, the last two librettos Smith devised for Herbert were for *Little Nemo* and *The Tattooed Man*, both tales of preposterous neverlands so dear to Smith. Indeed, only one of his librettos for Herbert, that for the 1905 *Miss Dolly Dollars*, was set in the contemporary and wholly Anglicized world that was home to its audiences.

If there were few truly first-class operetta librettists, and if the best of these were often doggedly conservative, there were still fewer great composers. Sousa offered a single work, one he had composed in a much earlier period. Walter Damrosch tried his hand at a score for an anti-war satire called *The Dove of Peace*. Luders turned increasingly to musical comedy. All three men created nothing but failures.

Ivan Caryll and Karl Hoschna most clearly understood what critics and playgoers preferred. But Hoschna wrote largely in a musical comedy idiom. Next to *Three Twins* his greatest success was *Madame Sherry*, and this "musical vaudeville" came as close to operetta as he would venture. Its story could have served operetta or musical comedy equally well. A young bachelor, supported by a rich uncle, has kept in his uncle's good graces by pretending to have a wife and several children. When his uncle suddenly appears, the young man must hastily hustle up a family. By the evening's end he has confessed his deception and agreed to marry a charming young lady, who happens to be his uncle's niece. The basic story, then, was sufficiently ambivalent to work as comic opera or musical comedy. Its treatment was similarly ambivalent, leaning more toward the use of the vernacular without subscribing to it as fully as did most of the era's more brash musical comedies.

In its final form, however, especially in the loose construction of its last act, *Madame Sherry* betrayed its affinity for turn-of-the-century musical comedy. The last act took place on the uncle's yacht, which remains anchored offshore until the uncle can determine the real relationship of the various principals. Obviously the situation could have been resolved in a few sentences, but that, of course, would have obviated the very need for the act. So the librettist, Otto Harbach, stretched it to the requisite length by conjuring up pretexts that allowed his principals to perform specialty routines or to sing interpolations.

Confronted with such generic imprecision, Hoschna composed a score that, Janus-like, looked both ways at once. Certainly the fervent "Birth Of Passion" would dignify any comic opera. But the show's great hit, "Every Little Movement," was a dainty piece, looked on in its day as a "polka française" and equally comfortable in either the period's operettas or musical comedies. Another hit from the show, a ragtime interpolation, "Put Your Arms Around Me, Honey," had an obvious musical comedy stamp. The rest of Hoschna's score swung back and forth, but if anything, remained closer to musical comedy.

Caryll is harder to characterize. On the one hand, his writing tended to be a little lighter than the Viennese, yet his graceful waltzes, his lively marches, and the sweet, flowing lines of so many of his other pieces marked them as belonging to the world of operetta. On the other hand, his years of writing musical comedies in England were almost as often evidenced by some of the gingerly tripping ditties, smacking inescapably of the music hall, which were interspersed with his more continental melodies. One can also discern his gradual Americanization as, in time, his waltzes assumed the narrower range and more homey sentimentality of his new countrymen.

Caryll's masterpiece is unquestionably *The Pink Lady,* and though, as we have said, it was customarily advertised as a musical comedy, its sweep and gaiety, its very essence, cry out that it is Viennese waltz-operetta at heart. Admittedly, its story, like that of *Madame Sherry* and numerous other contemporary pieces, could have been employed with ease in either musical comedy or comic opera. *The Pink Lady* skillfully interwove two simple tales, one recounting the exposure of a mysterious figure who has been stealing kisses from passing girls, the other of the hero's final, nostalgic fling with a lady of the demimonde before his wedding.

What placed the show squarely in operetta's corner was its fundamentally romantic outlook, its tightly knit construction, its excellent integration of music and story, and the lyric nature of the music. Caryll's great waltz from this work, a song known in the show both as "My Beautiful Lady" and as "The Kiss Waltz," lingers on as virtually his only composition still recognized today. But the remainder of the score was also excellent. Several other waltzes, notably "Love Is Divine," offered the same Middle-European feeling, even if they were not as memorably melodic as "My Beautiful Lady." "In A French Girl's Heart" was a pert French military march, a song that could easily have come from a classic opéra bouffe. Properly played, "By The Saskatchewan" would make a lovely barcarole. "Donny Didn't, Donny Did," which critics reported earned encore after encore as singers moved up and down staircases on the stage, was a superb concerted piece. If it lacked the grandeur of concerted pieces from earlier comic operas or from Herbert's later ones, it fit the tenor of the work like a glove. Bowing to modish necessity, Caryll began the final act with a contemporary two-step. Yet this, too, had a Continental flavor and the composer marked it accordingly, "Parisian Two-Step."

In his later successes during this period, *Oh! Oh! Delphine* and *The Little Café*, Caryll moved slowly toward lighter musical comedy forms. Even here, though, older voices seeped through. *Oh! Oh! Delphine*'s "Venus Waltz" and *The Little Café*'s "Thy Mouth Is A Rose" and "Just Because You're You" are Viennese in style. After the seven seasons following *The Merry Widow*'s premiere, Caryll composed increasingly within the era's accepted musical comedy formulas. His biggest hit, apart from "My Beautiful Lady," was a song for the 1917 musical comedy *Jack O' Lantern*. That song, "Wait Till The Cows Come Home," was a piece of cracker-barrel Americana without a taint of Old World influence.

Rudolf Friml appeared on the scene almost at the end of this period. Like virtually all his colleagues, he was an immigrant, a Czech who brought with him a special feeling for Middle-European art. Friml was fortunate. His first show, *The Firefly*, was devised as a vehicle for a great operatic soprano, Emma Trentini, and thus allowed Friml a free hand with the music. Otto Harbach was called in for the libretto. As one of the younger generation of librettists, he eschewed the yesteryears and Ruritanias that Smith or Stange might have selected and wrote instead a wholly contemporary story, set in the Manhattan and the vacationers' Bermuda of his day. His story dealt with an Italian waif who becomes a great opera singer, and thereby further justified the soaring score that Friml composed. (The conductor for the original production was Gaetano Merola, who later moved on to found the great San Francisco Opera.)

Probably because of Friml's music as well as Miss Trentini's presence, producer Oscar Hammerstein advertised the show with a rather singular label, not as a comic opera but as "a comedy opera." The term elevated it a notch or so, inching it away from ordinary playhouses and setting it on the doorstep of grand opera auditoriums. Yet at the same time the word "comedy" placed a stronger emphasis on the nature of the offering than the adjective "comic."

The musical was a triumph for Friml. In this respect, his debut was unique, for without exception his competitors had all begun their careers with failures. Naturally, Friml awarded the more fioritura material to Miss Trentini. She made popular the gorgeous "Giannina Mia" and stopped the show nightly, just before the finale, with a pyrotechnic showpiece, "The Dawn Of Love." She also sang several lighter pieces, including "Love Is Like A Firefly," "When A Maid Comes Knocking At Your Heart," and the trite march, "Tommy Atkins On A Dress Parade." The work's remaining popular hit, the lovely waltz "Sympathy," was assigned to her romantic vis-à-vis and her rival in the plot. Nor did Friml skimp when it came time to write for the comedians. The haunting "Something" was allotted them, despite its undercurrent of sadness.

Friml's only other work of this period, *High Jinks*, was much more of a musical comedy. Yet even here, the musical line of a waltz such as "Love's Own Kiss" is filled with Middle-European strains. For a decade after *High Jinks* Friml vacillated between pure operetta and musical comedy with operetta echoes. Not until a brief period beginning in 1924 did his natural musical bent seem to have truly free rein again.

The great composer who did hit his peak in this period was, of course, Victor Herbert. Eleven of his shows premiered during these seven seasons. Some of these, *Little Nemo, The Prima Donna, When Sweet Sixteen, The Duchess,* and an unrelated work, *The Madcap Duchess,* are scarcely remembered by anyone except aficionados. *Old Dutch* retains an important place in theatrical record books not by virtue of Herbert's score but because it marked the Broadway debut of Helen Hayes. Unfortunately, one superior score was dragged down by insuperable problems with its libretto. As *Algeria,* the "musical play" failed miserably in 1908. Revised and revived a year later as *The Rose of Algeria,* it failed again. Two of the score's best songs are still heard occasionally: the glowing "Rose Of The World" and the evocative "Twilight In Barakeesh." Yet even the work's secondary numbers, songs such as "Boule' Miche' " and "Love Is Like A Cigarette," represent Herbert at his most delightful.

One Herbert success, *The Enchantress,* is recalled today, if it is recalled at all, for the glorious paean Herbert gave its heroine for her first entrance, "The Land Of My Own Romance." While the song and the rest of the score sound now like any other good Herbert material, Herbert himself and many critics of the time saw it as a breakaway achievement. Herbert told an interviewer, "I determined, when I started its composition, to disregard absolutely every foreign impulse and to write in a frank, free American style." The *Dramatic Mirror,* exemplifying those who fell in line with Herbert's thinking, wrote that the score of *The Enchantress* "typifies American popular music, so far as we can be said to have any type. At least, it is far removed from the sprightly brilliance of the recent Viennese inundation." Herbert's biographer, with the advantage of forty years' distance, took a more balanced view. Waters wrote that Herbert's remarks

> show the trend of Herbert's efforts, and *The Enchantress* itself shows the mistaken notion he entertained of American style. The score did indeed have an extraordinary freshness and vitality; but musically it was European to the core . . . the final product was a combination of

> French and Austrian styles which wholly belied what Herbert said and thought he was trying to do. The music was so beautiful that nobody cared.

Librettos for Herbert's shows in this period represented a mixed bag. Three—*The Prima Donna, When Sweet Sixteen*, and *The Duchess*—at least in outline told stories of contemporary Western high society, much as so many of the better Viennese importations were doing. *The Madcap Duchess* told a similar tale, but set its action in the drapey elegance of the eighteenth century. *Old Dutch*, despite its name, had an up-to-date setting, but its comic complications centered on middle-class figures, and its music had less breadth and depth than Herbert at his best. For all practical purposes it was a musical comedy. *Little Nemo*, stemming as it did from a comic strip, and *The Lady of the Slipper*, based on a nursery tale, were special cases. Thus is it undoubtedly significant that Herbert's best scores among these lesser shows were for the two plays that told the most theatrically romantic adventures. *The Enchantress* was set in the Ruritanian principality of Zergovia, where a wicked minister attempts to disinherit his prince by marrying him to a commoner, a prima donna. When the singer turns out to have royal blood the minister is thwarted. *Algeria*, as its title hints, was still more romantic, touching on the allure of Arabian Nights and the mystique of the Foreign Legion. Its heroine, the Sultana Zoradie, searches for the unknown author of a poem that she has fallen in love with.

Curiously, Herbert's biggest popular success in this period, his only one to run more than two hundred performances, was another forgotten work, *The Lady of the Slipper*. This 1912 retelling of the Cinderella tale was an odd piece indeed. Conceived as a vehicle for a glittering array of performers—Elsie Janis, Dave Montgomery and Fred Stone, Irene and Vernon Castle, Joseph Cawthorn, none of whom was noted for great singing—it forced Herbert to write music confined to a much narrower range than was his wont. In one instance, with "A Little Girl At Home," the lone song to achieve some popularity in its own day, Herbert accomplished his end by writing a number with two musical parts. For Miss Janis he composed a repetitive, almost monotone, line, giving her better-voiced prince a more engaging melody. However, for the most part the challenge stymied Herbert, cramping his more open predilections. His best material was for the show's dances and pageantry, where singers' limited abilities could not restrict his inspiration. As for the rest, Waters rightly concludes that it has "little life of its own."

That leaves Herbert's two triumphs, his 1913 *Sweethearts* and his great masterpiece, the 1911 *Naughty Marietta*. Both were offered as "comic opera," a term by that time obsolescent. Given the merits of Herbert's scores for *Algeria* and *The Enchantress*, it is particularly telling that these two great scores were composed for the most romantic stories of all. Some might replace "romantic" with "old-fashioned," for there can be little argument that both told the sort of story Lillian Russell and her era had delighted in.

In *Sweethearts* we return to Graustark, this time named Zilania, for much of the action (the first half takes place in Bruges). The story revolves around a pretty young laundress, Sylvia, who, as only one man knows, is really the crown princess of Zilania. As a baby, she was left in a meadow near Bruges, found, and adopted by the laundry's proprietress, a woman known lovingly as Mother Goose. Now the man who left her in the field, a minister named Mikel Mikelovitz, is determined to relocate her and restore her to the throne. A motley collection of comic villains conspire to prevent her return. Sylvia makes matters difficult for everyone by falling in love with "a military lothario." Not until moments before the final curtain does she transfer her affections to Franz, currently heir presumptive of Zilania. They promise to rule Zilania together.

Waters has written of *Sweethearts*'s score, "From the opening chorus ('Iron! Iron! Iron!'), with its suggestion of a monotonous chore, to the closing finale, which exceptionally carried as much action as the first finale, the music exuded warmth and color, high spirits, romance, and occasionally serious contemplation." And Waters barely suggests the varied richness of Herbert's score, which moved from the cutely rhythmic "Jeanette And Her Little Wooden Shoes" to the radiant sanctity of "The Angelus" to the sturdy yet passionate "Every Lover Must Meet His Fate" to the homey "Cricket On The Hearth" to the uplifting "On Parade" to the jaunty "Pretty As A Picture." Capping all these and others was the irresistibly compelling title waltz. Not a song for a weak singer, Herbert undoubtably composed it knowing the great Christie MacDonald would introduce it. Its range is startling, just a half-step short of two octaves. Yet the soprano carried it off handsomely, as not just the period's critics could attest. MacDonald's primitive 1913 recording confirms what an incomparable voice Herbert was writing for. The song remains one of Herbert's best-loved, best-remembered melodies.

But *Naughty Marietta* has no fewer than five equally loved, equally remembered melodies. In this instance Herbert was composing for an even more fancifully romantic adventure—the story of a noble lady in disguise, of pirates, and of patriotic backwoods soldiers, all set in the lush beauty of eighteenth-century New Orleans. And he had not one, but two great opera singers to write for—the same Emma Trentini for whom Friml would later write *The Firefly*, and another member of Hammerstein's defunct opera troupe, Orville Harrold. Rida Johnson Young's plot recounted the difficulties a pretty countess meets when she runs away from an unhappy marriage and sails to America with a group of casquette girls. She is courted by Etienne Grandet, the son of New Orleans's governor. The treacherous Etienne leads a double life, harrassing his own city disguised as the buccaneer Bras Priqué. Etienne's nemesis is Captain Dick, who, along with his Rangers, wishes to rid America not merely of pirates but of foreign overlords as well.

Early on in the show, Dick and his men make their first appearance, announcing their patriotic determination in their stirring anthem, "Tramp! Tramp! Tramp!" Marietta reveals much about herself, too, in the wry, coy title song, one of Herbert's lesser-known delights. Later in the act, Adah, the beautiful mulatto whom Etienne has callously discarded, sings her sultry, poignant lament, " 'Neath The Southern Moon." Creating a stunning contrast, Herbert placed Marietta's brilliant showpiece, "Italian Street Song," back to back with Adah's somber soliloquy. Waters notes the problems these songs present. The joyous "Italian Street Song" confronts a singer with harrowing difficulties, requiring a glittering supple coloratura who can bring precision and purity to its skips and runs. Yet Adah's aria is almost as difficult, though it may sound deceptively simple. It calls for a contralto to sustain its long musical lines without sacrificing a moment's fervor.

Herbert saved his last two gems for the very end of the opera. Familiarity has dulled the edge of the chromatics with which Herbert began the chorus of Dick's emotional confession, "I'm Falling In Love With Someone." This is followed almost immediately by Herbert's most famous song, "Ah, Sweet Mystery Of Life." Although the song is played instrumentally earlier in the evening and the briefest snatches of it are sung several times, the whole song is not sung until seconds before the final curtain. It brings the work to a ravishing climax. While the other songs could not hope to be on a level with the score's outstanding melodies, they are very good indeed. (Gaetano Merola was *Naughty Marietta*'s first conductor, as he later was for *The Firefly*.)

If Herbert's conceptions of his music for *The Enchantress* were faulty, he nonetheless understood that his magnificent full-throated arias and the type of musical for which he composed them were passing from the scene. It was not entirely a matter of vogue, although that was undoubtedly a major consideration. With opportunities opening elsewhere—in vaudeville, cafés, and even silent films—youngsters were less and less willing to devote years to studying voice. Moreover, those who did and succeeded were reluctant to strain their voices singing the same thing eight times a week, week in and week out. Although enough fine singers remained available to the legitimate stage during the remainder of Herbert's life, his music, with a few exceptions such as *Eileen*, veered to the lighter requirements of musical comedy.

Hardly anything Herbert wrote at the end of his career could be deemed pure musical comedy, but a change in texture is incontestable. For example, compare his score for the 1919 musical comedy *Angel Face* with his score for the 1906 musical comedy. *The Red Mill*. Much of the latter's score is soaring and lyrical enough to allow the show to pass, on hasty glance, for operetta. This change was reflected in a perceptive and famous comment by Herbert. About the time of ASCAP's founding in 1914, he pointed to a young composer whom he predicted would "inherit my mantle." That composer was not Rudolf Friml, as might be expected, but a young man doing little more than interpolating songs in other men's musicals, Jerome Kern. Since the song that is supposed to have prompted the remark, "They Didn't Believe Me," is today generally accepted as the song that established the model for all future musical comedy writing, Herbert was clearly pessimistic about operetta as he understood it.

Had Herbert sat down with pencil and paper to analyze some statistics he might have had further cause for pessimism. If the rate of success of Viennese waltz-operetta was exceptional; if three out of five of the era's longest-running hits were Viennese operettas and a fourth was a first-class imitation; if in the long run (and this statistic would not have been available to Herbert) the music that has endured came from these Viennese works and their American cousins, still the picture was not all that encouraging. Moving beyond the first five hits—*The Merry Widow; The Pink Lady; The Chocolate Soldier; The Dollar Princess;* and *Three Twins*, the next seven—*Havana; The Girl Behind the Counter; Oh! Oh! Delphine; Alma, Where Do You Live?; The Lady of the Slipper; Madame Sherry;* and *High Jinks*—were either out-and-out musical comedies or borderline cases. True, three more Viennese operettas followed in close order—*The Spring Maid, The Rose Maid,* and *Little Boy Blue*—but beyond this top fifteen the picture changes radically, filled with nothing but the claptrap musical comedies and revues of the day. To list them all would be pointless. Who today remembers, for example, *Miss Innocence, A Knight for a Day, The Whirl of the World,* or *The Queen of the Moulin Rouge?* Not until well down in the top forty do operettas appear again with Kalman's *Sari,* De Koven's *The Beauty Spot, The Little Café, Naughty Marietta,* and *Sweethearts* bunched together. At the very bottom of the era's forty longest-running musicals are *The Count of Luxembourg* and *The Firefly*.

Of course their Broadway runs do not necessarily reflect either the intrinsic worth of many of these operettas or the values critics and playgoers of the day put on them. Many of the long-lived slapdash musicals ran as long as they did because of the charismatic appeal of their stars or the excellence of their mountings. Moreover, several of the better operettas, particularly Herbert's, appear to have been limited runs from the start, with post-Broadway bookings that no one chose to alter. Yet the fact remains that Americans seemed to prefer the giddy, gaudy, musical featherweights to the relatively more serious and romantic operettas they were offered.

Still, operettas might have continued to move along their own course, catering to and pleasing a large segment of American playgoers, but for one event thousands of miles from Broadway. In August 1914 the assassinations at Sarajevo in Serbia precipitated a world war. Almost immediately operetta was in trouble.

7

Interregnum: The Rise of the Revue and the Revitalization of Musical Comedy

For a time World War I might have been seen as not only the war to end all wars, but also as the war to end operettas. Ensuing years proved that the war ended neither wars nor operettas, but for a few seasons operetta was an innocent victim of the fighting, caught in a cross fire of international enmity. The battles had scarcely begun when many Americans began to view the Central Powers, the Austro-Hungarian Empire among them, as potential foes. Things Austrian or Hungarian first became suspect and then, after we entered on the side of the Allies, anathema. Both *Variety* and the *Dramatic Mirror* reported the hostility with which once-welcomed foreign confections were greeted. *Variety* suggested that the failures of two generally praised 1917 operettas, *Rambler Rose* and *The Riviera Girl,* stemmed from playgoers' perception of them as Viennese, although actually only the latter was. Two months later, Franz Lehar's *The Star Gazer* was literally hooted off the stage, and those hoots attested as much to America's strident anti-Austrian sentiments as to the operetta's many weaknesses.

But it was more than America's refusal to put dollars into enemy pockets that sent operetta into a momentary retreat. The warring European nations had called up many of their younger writers for military service. Their older writers either had peaked or were thrown off balance by the turmoil around them. In truth, few operettas of real merit premiered on German or Austrian boards during the war. For the first time in many years, American writers and producers found they had the field virtually to themselves. They took brilliant advantage of the opening.

Most of these writers and producers were American-born or, like Irving Berlin, had at least spent important formative years here. If their backgrounds and training did not totally alienate them from the Continental forms that had dominated operetta, it prejudiced them heavily in favor of native ideas and idioms. They preferred to offer their time and talent to forms that seemed to them more homegrown, and so they turned their attention not to creating American operettas so much as American revues and musical comedies.

Florenz Ziegfeld was the presiding genius in the revue field. His first *Follies* had premiered in New York in 1907, the very same season that celebrated the premiere of *The Merry Widow*. Though his original inspiration had been French, Ziegfeld nevertheless moved his yearly editions of the show along lines determined by his own unique taste. In 1915 he hired Austrian immigrant Joseph Urban to design all his future *Follies*. Together Urban and Ziegfeld brought to the revue form an imaginative, tasteful opulence and grandeur such as Broadway had never seen before nor, arguably, would see again. While his line of beautiful, gorgeously gowned girls was a principal attraction, Ziegfeld was careful to fill the rest of the show with great clowns spoofing the lunacies of the day and with songs he hoped would have instant popularity.

Ziegfeld's purported indifference to his songwriters is legendary, yet he clearly was interested enough in his scores to establish an implicit policy for his underlings. With the exception of Irving Berlin, no major composer was ever signed to compose a basic score for the *Follies*, but Ziegfeld and his associates regularly called on other fine writers to interpolate songs. Ziegfeld's demands were small. He wanted simple, catchy tunes to set feet tapping and a few, more languid melodies for his girls to parade to. An occasional short ballet or pretentious tableau gave Victor Herbert or Rudolf Friml a better chance to express themselves, but such moments were exceptional. Other producers quickly imitated, or tried to imitate, Ziegfeld's success. As a result, the decade following 1915 remains the heyday of the extravagant revue.

Developments in musical comedy were more important and longer lasting. From the time of the first revue on Broadway in 1894, many a musical comedy was mistaken for a revue and vice versa. Though such confusion might seem startling from nearly a century's remove, it is understandable given the nature of our early musicals. A distressing number of early musical comedies had the slenderest of plots—as often as not simply a tour of some place—and quite often even the slim connecting threads were lost halfway through the show. To fill out the two and a half hours of entertainment audiences expected, all manner of extraneous specialty acts were brought in. The result was an evening of shreds and patches. At the same time, virtually every early revue tied together its sketches and musical numbers with a tenuous story line, thereby conveying something of a musical comedy air. No wonder, then, that critics and playgoers were often both confused and contemptuous. Of course, the better musicals, from those of Harrigan and Hart onward, were strongly plotted. But these better shows constituted little more than a majority of the entertainments that called themselves musical comedies.

Matters began to change permanently for the better in 1915 when a producer named Ray Comstock and an agent named Elisabeth Marbury asked Jerome Kern and Guy Bolton to write a musical for the Princess, a tiny, failing theatre on 39th Street in New York. All they were seeking was an "Americanization" of an English musical, with Bolton to do a minimum of revising and Kern to interpolate a handful of new melodies, just as he had done so successfully the year before in *The Girl from Utah*. But an early rehearsal warned everyone the show was unacceptable. Bolton set about making far more extensive alterations, while Kern contributed the main body of songs, with additional American numbers added to complete the score. Not a single song from the original English show survived the rewriting.

That first Princess Theatre show, *Nobody Home*, was little more than a modest success. But it was followed quickly by *Very Good Eddie*, and, after P. G. Wodehouse joined Bolton and Kern, by *Oh, Boy!* and *Oh, Lady! Lady!!*, along with two other shows by the trio that played houses other than the Princess, *Have a Heart* and *Leave It to Jane*.

These shows were not revolutionary departures from the past. If their stories ignored nobles and soldiers and grand prima donnas and instead looked with affectionate humor on the often comic behavior of everyday Americans, so had countless musical comedy librettos before them. However, Bolton did try to abolish jokes for jokes' sake. He tried, and generally succeeded, to make every line pertinent to the plot, moving it onward. Songs were carefully incorporated; they too usually helped further the plot's development. This was also not new. George M. Cohan attempted much the same in his better shows. But Wodehouse's lyrics were the best the American musical theatre had yet produced—comfortably colloquial, stylish and supple, literate and witty—and Kern provided sweet, curvaceous, easily singable and often unforgettable melodies for them. Magic melodies, sophisticated lyrics, and thoughtful books fell together in the most artfully crafted musical comedies America had yet created. The Princess Theatre shows were carefully polished gems. They boasted a unity of style and tone rare until then on our lighter lyric stage.

The Princess Theatre shows were not revolutionary, nor did they spark a revolution. The books of all too many musical comedies in the decades that followed were still claptrap—weakly plotted and embarrassingly dependent on Joe Miller jokes. But conscientious writers thereafter had models, and demanding critics a set of exemplars. Musical comedy did in fact improve, although not as rapidly or as totally as might have been hoped.

Revues and musical comedies alike also benefited from the commercialization of jazz, a fresh musical idiom that evolved out of ragtime and, in effect, superseded it. Urgent, dynamic, and informal when compared to most other show music, its metallic tone caught the brash spirit emerging not merely in the most stylish revues and musical comedies but in the whole nation. Much of it was blaringly unsentimental. When, in the form of the blues, it did turn sentimental, it effected a world-weariness that became increasingly fashionable.

Yet the rage for revues, the wartime success of the Princess Theatre shows, and the new vogue of jazz could not wholly stifle operetta. While Americans balked at importations from enemy stages, they still welcomed native efforts. The *Dramatic Mirror* occasionally underscored this receptivity in its own curious way. In July 1915, foreseeing the probable severance of ties with Vienna and Berlin, the paper urged Americans to write "comic operas" of their own. And it suggested that for subject matter they turn to nothing less than the war then raging overseas. This romantic notion of war, perhaps even the employment of the term "comic opera," suggested how irretrievably out of touch with the times the tradesheet was. Yet its desire to see an older genre flourish alongside newer competitors was not merely understandable but broad-minded.

Two years later, in March 1917, the paper carried a far longer article under the byline of Carl Wilmore. Although the article was entitled simply "American Operetta," its subheading indicated its real gist: "The Need Is for Intelligent Writers Who Take Their Work Seriously—Men of Talent Turning to Other Fields." Wilmore bewailed that even graduates of Professor Baker's celebrated course on playwriting at Harvard University could not see differences among the material for Ziegfeld's *Follies*, the pieced-together musical comedies of the day, and the well-structured plots of the better operettas. He rued, "They have a peculiarly American lack of the feeling for what is in good taste and possesses that elusive something which, for want of a better word, I like to call 'style.' " Echoing a by then familiar cry, he continued, "Only in Europe is the libretto recognized for what it is, a dignified form of the drama." He acknowledged that styles in playwriting change as much as styles elsewhere, and that by 1917 no one could seriously consider writing in Gilbert's mode or in the mode of an early, above-average American contribution, of which he saw *Robin Hood* as an example. Nonetheless, he argued, reasonableness and intellect could be brought to the modern libretto.

Wilmore neglected to comment directly on fashions in subject matter, apparently deeming treatment the only important issue. But at this point he came to what he surely considered the nub of his argument, "the secret of the good operetta libretto [is that] nothing tragic and nothing deadly serious is ever attempted. Even in the most dramatic situations you still remember, and feel, that it is operetta, and that the writer has kept quite within his métier." Perhaps belaboring a point he had already underscored, he went on, "The foreigners do it all better," and then added the somewhat cryptic observation, "They write pungent, modern comedies with music—and that constitutes the modern operetta. They don't force things, they don't even try to be funny, for they understand that in operetta, as in comedy, this is fatal."

Thus, by indirection, Wilmore appeared to be advocating operetta in the contemporary Viennese pattern, putting aside the more exotic, arch romanticism and the more patent absurdities of older comic opera. In short, he appeared to be falling in line with the trend to move operetta closer to musical comedy.

When Wilmore finally touched on music, his notions were consistent with his "up-to-date" views of librettos. He had little use for what he saw as the grandiloquence of the earlier school. He deemed soaring grand finales, recitatives, and an overuse of large choruses as hopelessly antiquated. Some contemporary composers, he concluded ruefully, had not understood the need for trimmer musical lines:

> Mr. Friml, although schooled abroad, has not yet learned this; Mr. Herbert does better, only occasionally falling into this error. Franz Lehar, best of all modern writers, produces operetta that charms, thrills, and appeals, because he tries to remember that he is writing a play with music, not a lot of music to be fitted with words.

Musically, at least, Wilmore would have been pleased with the future course of operetta in America. Far and away the biggest American operetta success during the months in which we joined the war was *Maytime*, which opened August 16, 1917. The adjective "American" has to be employed advisedly. The Shuberts had optioned Walter Kollo's Berlin success *Wie einst im Mai*. But shortly before the show went into rehearsal, their production of Oscar Straus's *My Lady's Glove* was assailed as much for its Austrian origins as for its inherent weaknesses. Even switching the locale of the show's action to France and distributing free chocolates at each performance could not lure playgoers. (And one night some of the few who did attend returned the Shuberts' calculated generosity by distributing free, if overripe, tomatoes to the cast.) So in the end the Shuberts dropped all mention of *Wie einst im Mai*'s German beginnings in their announcements and set about to totally "Americanize" the show.

While the basic story outline was retained, the plot was reset in America, and every bit of Kollo's original score was discarded. Rida Johnson Young, a Baltimore beauty who had supplied Herbert with the libretto for *Naughty Marietta* and who would give Friml the libretto for a modest hit called *Sometime*, retold the tale of two ill-starred lovers who are never allowed to marry but whose grandchildren do. The action begins in the first half of the nineteenth century. The heroine is a rich, socially prominent New Yorker, the hero a poor struggling contractor. Their plans for marriage are thwarted by the young lady's snobbish father. As the acts leap across the decades, the lady is forced to marry a man who proves a ne'er-do-well and eventually leaves her penniless. When her home is put on the auction block, the contractor, now rich, buys it back for her. In the fourth and final act, set in the teens of the twentieth century, both the original lovers are dead. But the hero's grandson, over the objections of his rich, socially prominent and snobbish parents, weds the heroine's struggling granddaughter. With verisimilitude perhaps thrown to the wind, the grandchildren were portrayed by the same actress and actor who had played the first pair of lovers.

For the score, the Shuberts called on Sigmund Romberg, a Hungarian immigrant who had been churning out Tin Pan Alley ditties for Shubert revues. Romberg had first called attention to himself three years earlier when he had been permitted to set aside hurdy-gurdy rags and interpolate appropriate melodies into another Shubert importation, *The Blue Paradise*. His natural melodic bent was given free rein, and the result was his first durable classic, "Auf Wiedersehn."

Reveling even more in the freedom the Shuberts granted him for *Maytime*, he composed one of his greatest scores. The music was largely Viennese, tempered throughout with subtle American softenings. Moreover, recitatives, extended choral passages, and overflowing finales were minimized. "Will You Remember?" (often referred to as "Sweetheart") was, of course, the runaway hit of the evening. But two other waltzes from the show, "The Road To Paradise" and "Gypsy Song" ("Oh, Come Away With Me") were also popular and are still heard occasionally. Because the story spanned three quarters of a century, Romberg had a field day with period pieces such as a classic nineteenth-century mazurka and a twentieth-century rag as well as a salute to American minstrelsy in "Jump, Jim Crow," always a showstopper in performances. The variety, color, and appealing melodicism of Romberg's score coupled with the unapologetic sentimentality of Mrs. Young's libretto appealed to a war-weary nation. *Maytime* was reputed to be the first choice of soldiers in New York awaiting embarkation for the front. Up to that point, then, Americans shared Wilmore's criteria.

Maytime represented a departure from other offerings in another way, one that might have given Wilmore a moment's thought. Its plot did not unite its lovers happily ever after, though the conclusion of *Maytime* could be seen as a compromise. Unhappy endings were a rarity in operetta. They were not totally unknown, as demonstrated by, say, Lehar's *Gypsy Love* or Gilbert and Sullivan's *The Yeoman of the Guard*, but they were, for the most part, left to grand opera. Nevertheless, while they did not become commonplaces, bittersweet finales would hereafter appear with more frequency in operetta, and no one would advocate or be inspired by them as much as Romberg.

For all *Maytime*'s success, operetta still had an uphill battle in the years following the war. The Cinderella Era in musical comedy had begun. For several consecutive seasons the poor-little-girl-makes-good plot seemingly monopolized musical comedy. As often as not the girl was a hard-working secretary, frequently Irish, who ended by marrying her boss's son. A common variation was the poor girl who rose to be a star. This was not the world of the rich, selfish dollar princess. In approach and in tone these Cinderella musicals were thoroughly middle class. The popularity of this motif, coupled with the peak of revues' popularity and the advent of theatrical jazz, left little room for Old World plots swept along by graceful waltzes. Of course, there were some notable exceptions.

In 1915 Friml's *Katinka* was a major hit, giving us "Allah's Holiday." Four years later, in 1919, Victor Jacobi, who was by then an American citizen, and famous violinist Fritz Kreisler, who was not, combined their talents to create a captivating score for a plot that would have done Viennese librettists proud. The result was *Apple Blossoms*. The plot revolved about a young man and young woman who grudgingly marry at the insistence of their rich uncle. Believing that they have made a grave mistake and are hopelessly incompatible, the newlyweds agree they are free to resume their old flirtations. At a masked ball they unwittingly fall in love with each other. The plaintive waltz, "You Are Free," was the reigning hit of the show, but a second waltz, "Who Can Tell?," became popular years later with a new lyric as "Stars In Your Eyes." The hero's lively "Little Girls, Good Bye," with its insistent "I love the girls, girls, girls, just the same," evoked happy memories of *The Merry Widow*'s "Ja, das Studium der Weiber ist schwer" and its chorus of "Weib, Weib, Weib."

Two years after *Apple Blossoms*, Sigmund Romberg reworked Franz Schubert melodies and set them in a tale of an unfulfilled lover. That lover purported to be none other than Schubert himself, who loses his beloved Mitzi through a case of mistaken identity. Only as Schubert lies dying in the last act does Mitzi belatedly realize her error. No one cared that *Blossom Time* had originally been a Middle-European success called *Das Dreimädlerhaus*, or that the rest of Europe had embraced it under one title or another. In all its redactions the operetta gave Schubert a fictitious, unfulfilled romance. But the American version jettisoned most of the European score, selecting different Schubert melodies for revision. Romberg's most successful adaptation was "Song Of Love," a waltz developed from a principal theme of the first movement of the *Unfinished Symphony*. From a later passage in the same work Romberg derived "Tell Me, Daisy." Incorporated with minimal alterations were Schubert's famed "Serenade" and "Ave Maria." Later hits came and went, but *Blossom Time* toured American stages for the next quarter-century.

A few lesser hits, *The Lady in Ermine* for example, could be cited. This show was an importation, substantially revised and larded with interpolations by Al Goodman and Sigmund Romberg. The latter's

"When Hearts Are Young" helped the entertainment lure playgoers. But the list of noteworthy and successful operettas was skimpy indeed. By 1923 or early 1924 cogent arguments could have been mustered to prove operetta had seen its day.

8

The Flowering of Traditional American Operetta: Friml and Romberg

"The history of musical comedy has passed through a variety of phases, but the type that persists, that shows the signs of ultimate victory, is the operetta—the musical play with music and plot welded together in skillful cohesion. These are the only kind that are revived years after their first presentation. Other forms of musical entertainment have brief vogues and crowd out the operetta for a while. But the good old singing show always comes back, each time stronger than ever."

In later years the author of these lines, Oscar Hammerstein II, would think twice about using the term "operetta," but he never deviated from his ambition to realize a lyric stage given over to musical plays that intelligently integrated music and plot. Twice—at separate times nearly twenty years apart—his librettos and lyrics ensured that operetta's popularity would be rekindled. The first renaissance lasted a mere five years and represented the last glorious heyday of traditional operetta. The second revival lasted for decades and was in some ways so revolutionary that many refused to admit it dealt with operetta at all.

Hammerstein's 1925 essay was prompted by the success of *Rose-Marie,* his collaboration with Otto Harbach, Herbert Stothart, and Rudolf Friml, which opened on Broadway in the late summer of 1924. The show was almost as much a breakaway as a success. While any number of operettas before *Rose-Marie* had been set in America, their stories, with rare exceptions, were such that they could have just as easily been played out against European backgrounds. But *Rose-Marie*'s North American setting—the Canadian Rockies—was an integral part of its story. The mountain hideaway of the hero, the Indian encampments, and the very relationship of white men and red men figured inextricably in the plot. Indeed, the pivotal romance between the white hero and the part-Indian heroine injected, however implicitly, a new racial element into the stock comic opera motif of love between social unequals. (Indians had figured in American musicals as far back as the 1794 *Tammany* and in recognizable modern comic opera in the 1894 production by the Bostonians of *The Ogallallas.* But these had not been especially successful.) In *Rose-Marie* the fur trapper, Jim Kenyon, loves the half-breed singer, Rose-Marie La Flamme, but their romance hits a seemingly insurmountable snag when Rose-Marie is led to believe Jim has murdered an Indian. The Mounties arrive to help unmask the real killer, and the lovers are finally united.

Even Friml's music represented certain departures. While there was little that was genuinely Indian in the musical lines or chromatics of either the impassioned "Indian Love Call," with its diminished progressions, or the throbbing "Totem Tom Tom," both marked unique changes from traditional Middle-European musical mannerisms, and both somehow managed to convey a suggestion of Indian primitivism. Most of the show's other songs, including the great title number, were not so markedly novel. They could have come from almost any Friml operetta.

Basking in the work's popularity, Hammerstein observed,

> There has been no change in the theatrical fashion so swift and complete as that which has taken place since *Rose-Marie* pioneered its way to Broadway last September. Here was a musical show with a melodramatic plot and a cast of players who were called upon to actually sing

> the music—*sing*, mind you—not just talk through the lyrics and then go into their dance.

Hammerstein had good cause to be proud. *Rose-Marie* was the biggest musical success of the 1920s, triumphing not just in New York—with 557 performances—and across America, but in virtually every major Western theatrical capital. Its London run of 851 showings far surpassed its New York stand, while in Paris it established a long-run record of 1,250 repetitions.

Yet, in claiming no show ever effected "so swift and complete" a change, Hammerstein had to be resorting to hyperbole. Although he gave knowing attention in his article both to Gilbert and Sullivan and to Viennese waltz-operettas, he ignored the phenomenal success and influence of *H.M.S. Pinafore*, which had arrived just forty-five years before and thus could have been alive in the memory of any number of theatrical figures, and that of *The Merry Widow*, which had premiered in New York some seventeen years earlier. *Pinafore* was almost incontestably the most influential musical ever to reach New York, while *The Merry Widow*'s importance, despite statistical evidence, was every bit as great as *Rose-Marie*'s.

Of course, both earlier musicals had been importations. In *Pinafore*'s day America had no librettists, lyricists, or composers equal to the task of immediately bringing out domestic counterparts. It took American writers several years to master the art of comic opera. Nonetheless, the onslaught of claptrap musicals on Broadway in the season following *Pinafore*'s premiere marked the real beginning of America's love affair with popular musical theatre. *Pinafore* had single-handedly opened all of America's stages to song and dance. By the time *The Merry Widow* appeared, there were any number of Americans capable of providing almost instant domestic responses, even if several of the most productive composers were adopted sons. True, their response reembraced an older style of comic opera rather than slavishly imitating the newer waltz-operettas. But regardless of stylistic allegiances, American operetta did quickly compete with Viennese. For audiences demanding the genuine Viennese article, there were producers ready and willing to import as many as the public demanded.

To bolster his argument, Hammerstein cited the reception accorded to *The Student Prince* and *The Love Song*, which opened in the same season as *Rose-Marie*. *The Love Song* was a Shubert importation that Harry B. Smith had adapted to American tastes. Belonging to the *Blossom Time* school of operetta, it used Edward Kunneke's adaptations of Offenbach's melodies to decorate a fictitious life of the French composer. For *The Student Prince*, Romberg set to music Richard Mansfield's vehicle, *Old Heidelberg*, and was able to find inspiration in the very traditional material. The plot recounted a foredoomed romance between a prince and a beer-garden waitress, played out in glittering courts and an autumn-hued German college town. An argument could be made that the story was not totally traditional: the saddened but essentially unrebellious lovers accept the social dictates of their world. The heroine is not discovered to have royal blood, nor is there any hint that future generations will resolve their woes. There is, instead, a touch of fatalism rare in pieces of the period. Against this background Romberg offered a roof-raising drinking song, a moonlit serenade, and, in the operetta's principal waltz-duet, "Deep In My Heart, Dear," an almost classic Viennese waltz, moving from theme to theme, with each theme varying in tempo and mood. Only the fact that the first of the song's themes was in 4/4 time kept it from being truly classic in form. Minor deviations such as this apart, *The Student Prince*, unlike *Rose-Marie*, adhered tenaciously to all of operetta's time-tested formulas.

The Student Prince did have something in common with *Rose-Marie*. Unlike *Pinafore* and *The Merry Widow*, both were American-made. And American offerings had rekindled interest in operetta in the mid-twenties on Broadway. This was one point that for some reason Hammerstein overlooked. Both shows also had scores by immigrant composers, a fact which colored not only their music but Americans' attitudes toward it.

Yet while a large number of operettas followed on the heels of these shows for the remainder of the decade, their number attested as much to Broadway's prodigality throughout the twenties as to any special demand for operettas. Indeed, as always, the number of musical comedies and revues continued to exceed the count of operettas. The popularity of the extravagant revue was undoubtedly waning, but musical comedy was retaining its unbreakable hold on public affection. Hammerstein's perception of the swiftness with which operettas succeeded one another is therefore open to dispute, and his contention that operettas precipitated a "complete change" is wholly untenable.

Moreover, a major transformation had been slowly taking place in musical comedy. Three months after *Rose-Marie* opened and just one night before *The Student Prince* premiered, *Lady, Be Good!* arrived to confirm and complete that transformation. Its Gershwin score proclaimed that thereafter the most desirable musical idiom for musical comedy was jazz—American-bred jazz, with its singular, easily identifiable harmonies and rhythms. Music and librettos alike would accentuate the distinction between musical comedies and operettas. As a result, earlier confusions evaporated, and the two genres would rarely overlap again. There would be exceptions. Romberg would still talk of musical comedy when he meant operetta, and one of his best operettas would reach New York labeled "a romantic musical comedy." However, for the most part this newfound distinction applied.

The clear-cut separation had advantages and disadvantages. Playgoers had a better idea of the nature of the entertainments for which they were buying tickets; writers could strive for tonal unity and thereby work comfortably in their natural styles. (Here, too, there would be interesting exceptions. As we shall see, in the very next season Gershwin himself tried his hand, not unsuccessfully, at operetta.)

Yet while such marked separation was helpful, it had the disadvantage of corralling advocates into opposing camps. Probably no one was surprised when Sigmund Romberg opted for operetta in 1925. He was a warm, kindly man, and his comments were couched in the most courteous terms:

> Casting no reflection, you will notice today that there are types of musical comedies which "flop," in the parlance of the theatre. On the other hand, you will find one here and there that plays for a seemingly endless time to capacity houses. The latter, if you will observe, is usually the musical comedy which leans toward the operatic rather than the jazz type.

Romberg christened these works "light comedy opera," but he clearly meant operetta. He felt the reason for this favoring of operetta derived from "people's appreciation of music [being] raised quite a bit in the last four years. They have learned to like the better class of music." A generous man, Romberg attributed this improvement to an often ignored competitor: ". . . it is to the movies that we owe this sense of appreciation. With the advent of the big motion-picture theatre we also saw the augmented symphonic orchestra. This played real music and the movie-going public learned to like it."

What may have been a more surprising broadside on jazz musicals was launched by Jerome Kern. His attack was less temperate than Romberg's, but to his credit he backed his words with action, although it sometimes cost him dearly. He refused to release his highly praised songs from *Sitting Pretty* for public performance, lest they be manhandled by jazz orchestras. The failure of the public to hear these songs most likely hastened *Sitting Pretty*'s demise. Of course, Kern was assailing jazz orchestras and the public conception of jazz more than jazz itself. He was one of the earliest composers to cull from jazz patterns, even introducing the saxophone into pit bands, and he wrote what could pass as jazz material off and on throughout his career, albeit his heart was probably elsewhere.

On the other side, modernists gleefully had fun at operetta's expense. Many a 1920s revue contained a skit satirizing comic opera's most marked absurdities. The Algonquin coterie's revue, *The 49ers*, included just such a skit. The progressive Theatre Guild's *Garrick Gaieties* offered a more famous spoof, "Rose of Arizona," from the pens of none other than Rodgers and Hart. For the last half of the twenties neither side's thrusts had much effect. Operetta and musical comedy flourished next door to each other.

One reason for operetta's successful coexistence with more modern musicals was the welcoming or at least tolerant stance of most critics. Almost to a man they bowed and scraped joyfully at *Rose-Marie*'s appearance, although, if one reads between the lines, one gets the impression that they were slightly baffled about where to place the musical, despite its programs and advertisements, which distinctly labeled it "a musical play." Its sheet music employed the same term, although critics could not have been expected to have the printed songs in front of them. Deems Taylor and Heywood Broun, writing independently in the New York *World*, both called the show a "musical comedy."

Charles Belmont Davis, the New York *Herald Tribune*'s critic, saw it as a mélange of "drama, melodrama, musical comedy, grand opera and opera comique," while the *New York American*'s Alan Dale felt it moved the popular musical stage far along "on the road to grand opera."

However easy it is to belittle such confusion in retrospect, it becomes more or less understandable if one places oneself in the critics' chairs. By 1924 operetta had, after all, been somewhat déclassé for a number of seasons. Moreover, as we shall see in a minute, Friml had been moving clearly in the direction of musical comedy, while Hammerstein had written only musical comedy librettos to date and Harbach had not done an operetta libretto since 1917. If some reviewers acknowledged that Friml was returning to his more traditional, eloquent style, none took note of how close *Rose-Marie*'s plot was to that of older shows such as *Naughty Marietta.*

By the time *The Student Prince* opened a few months later, a proper sense of perspective had been restored. The *World* hailed it as "one of the finest, most robust and most stirring of all American-made light operas." The paper's sister, the *Evening World*, called the show "a Real Operetta" and proclaimed it "a delight." At the same time it couched some of its pleasures in words that almost paraphrased Hammerstein and foreshadowed problems: "It is old-fashioned—like Rose Marie for instance—in that the singers . . . could really sing."

As the parade of operettas continued, some reaction soon set in. The story for Friml's 1925 hit, *The Vagabond King*, was derived from Justin McCarthy's *If I Were King*. Its plot made the rabble-rousing poet François Villon ruler of France for a day and allowed him to vanquish Louis XI's enemies. The King rewards Villon with the hand of the beautiful Katherine de Vaucelles. Friml's score, his greatest, was jam-packed with still popular gems such as "The Song Of The Vagabonds," "Only A Rose," "Someday," "Huguette's Waltz," and "Love Me Tonight." The reviewers for the *Herald Tribune* and for the *World*, while praising *The Vagabond King*, did so in terms clearly meant to set themselves at arm's length from their readers and from the operettas they knew many of those readers admired. Thus the *Herald Tribune* predicted "for playgoers who love operetta . . . it is

Picture Gallery of Operetta in America

A scene from the 1911 revival of *H.M.S. Pinafore*. Marie Cahill is seen left of center as Buttercup, and Henry E. Dixey, center stage, as Sir Joseph Porter. (Theatre and Music Collection, Museum of the City of New York)

Jessie Bartlett Davis in her famous trouser role of Alan-a-Dale points a finger at Henry Clay Barnabee's Sheriff of Nottingham in the original production of *Robin Hood*. (Theatre and Music Collection, Museum of the City of New York)

Fritzi Scheff and Walter Pruette in Victor Herbert's *Mlle. Modiste*. (Theatre and Music Collection, Museum of the City of New York)

Lillian Russell, the most glamorous of all nineteenth-century prima donnas. (Theatre and Music Collection, Museum of the City of New York)

The principals and chorus in an ensemble from the original production of *Robin Hood*. Note the essential simplicity of the sets. (Theatre and Music Collection, Museum of the City of New York)

Left, Donald Brian as Prince Danilo inviting Ethel Jackson as Sonja, the merry widow, to waltz. (Theatre and Music Collection, Museum of the City of New York.) Right, De Wolf Hopper as Howdja Dhu gazes skyward in the original production of *The Begum*. (Theatre and Music Collection, Museum of the City of New York)

A view of the set and company of a 1908 road version of *The Merry Widow*. Again note the basic simplicity of the set. Details were achieved by careful, elaborate painting. (Theatre and Music Collection, Museum of the City of New York)

Left, Emma Trentini (star of *Naughty Marietta*, *The Firefly*, and *The Peasant Girl*). (Theatre and Music Collection, Museum of the City of New York.) Right, Evelyn Herbert (star of *Princess Flavia*, *My Maryland*, *The New Moon*, and revival of *Bitter Sweet*). (Theatre and Music Collection, Museum of the City of New York)

Twelve of the forty chorus girls who dressed as totem poles to sing "Totem Tom Tom" in *Rose-Marie*. (Theatre and Music Collection, Museum of the City of New York)

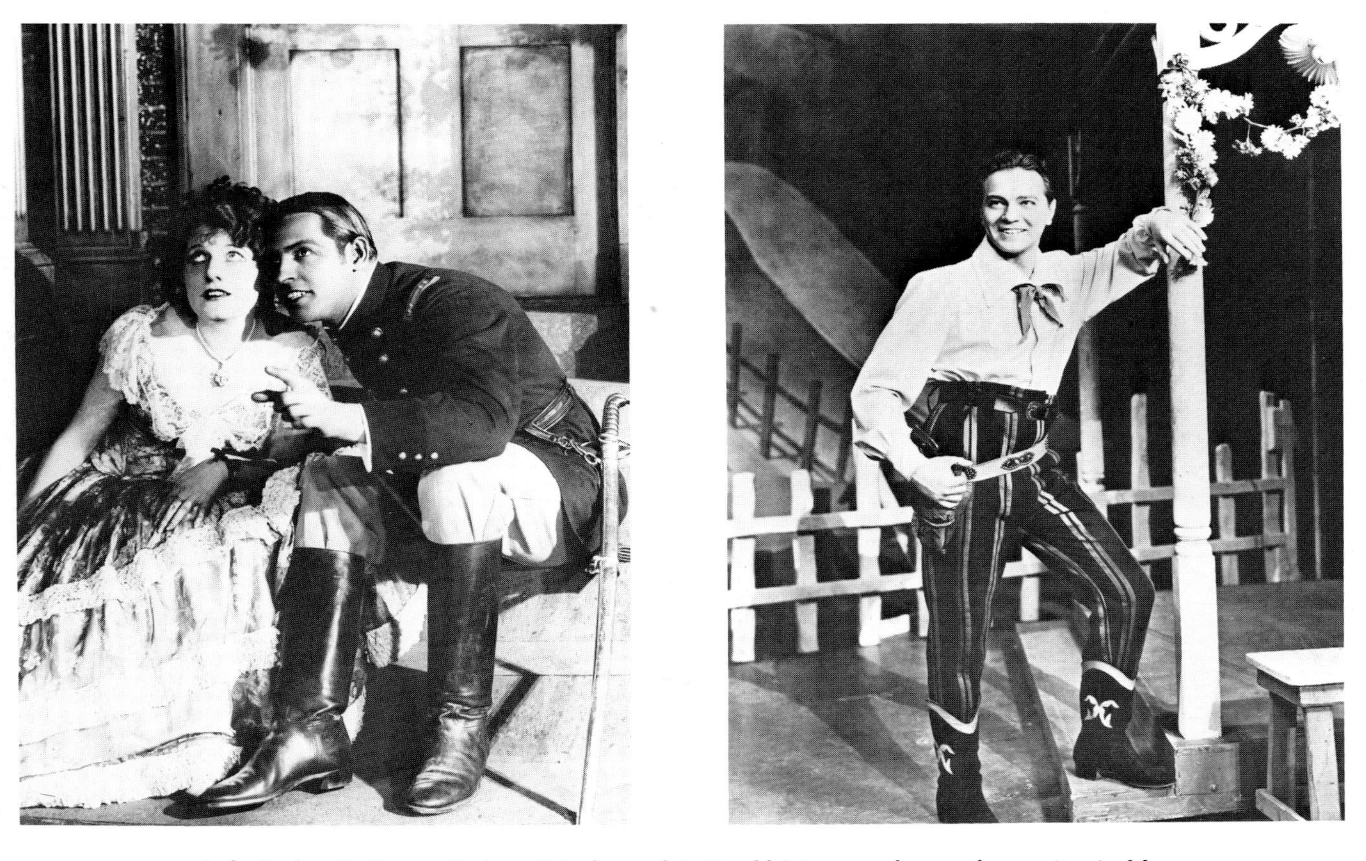

Left, Evelyn Herbert as Barbara Frietchie and J. Harold Murray as her northern suitor in *My Maryland*. (Theatre and Music Collection, Museum of the City of New York.) Right, Alfred Drake as Curly in *Oklahoma!*. (Theatre and Music Collection, Museum of the City of New York)

An Agnes de Mille dance integrated into the story of *Oklahoma!*. (Theatre and Music Collection, Museum of the City of New York)

Ezio Pinza and Mary Martin's reconciliation watched by the young performers who played Pinza's children in *South Pacific*. (Theatre and Music Collection, Museum of the City of New York)

Yul Brynner and Gertrude Lawrence in the original *The King and I*. The picture gives a hint of changing set construction and lighting when compared to photographs of *Robin Hood* and *The Merry Widow*. (Theatre and Music Collection, Museum of the City of New York)

Peers and Peeresses assembled for "The Ascot Gavotte" in *My Fair Lady*. (Theatre and Music Collection, Museum of the City of New York)

Zero Mostel as Tevya in *Fiddler on the Roof*. (Theatre and Music Collection, Museum of the City of New York)

our guess that this pleasure will be at their disposal for a long, long time." A trifle facetiously, the *World* concluded,

> There were sword play and vast drinking in the tavern, and, as is too customary in such matters, gadzooks and odds bodkins served for comedy. . . . Last season the population approved fervently of "The Student Prince," "The Love Song," and earlier, with equal intensity, of "Blossom Time." "The Vagabond King" is a graduate of the same school of entertainment, decidedly a graduate with honor.

Yet even this muted disdain melted if a critic enjoyed a particular operetta. When Davis, who seemed to do a slight sidestep in his review of *The Vagabond King*, fell in love with *Princess Flavia*, he praised it for proffering "a new standard of operetta."

For the most part, then, operettas were judged during the twenties not as a genre, especially not as a fustian, outdated genre, but as individual offerings, with or without their own merits. Yet occasionally a critic did admit to doubts about the genre. Alexander Woollcott saw *Countess Maritza* as "an unstinted effort directed at the vast audience which adored 'The Student Prince,' " only to continue, "If you enjoyed that interminable diversion, you should be rapturous at 'Countess Maritza.' It is not the kind of entertainment which your correspondent relishes ever." Over at the *Times*, its new critic J. Brooks Atkinson apparently felt the same way, although for the time being he withheld his strongest feelings. In his review of *Countess Maritza* he elected to do some discreet fence-sitting, informing his readers the musical was "a pleasing operetta of the old school." His sentiments clearly had not yet totally crystallized. In later years, when they had, Atkinson would become one of operetta's most vocal and vituperative enemies.

Critical even-handedness eventually began to wear thin, and as the twenties moved on, sides were frequently drawn. Perhaps surprisingly, much of the most forceful advocacy in morning-after reviews was in favor of operetta and against the excesses of jazz musicals. When what proved to be the last major operetta hit of the twenties, *The New Moon*, opened in late 1928, it brought to the fore not only staunch defenses by proponents of operetta, but some interesting confessions by critics.

The show offered a stellar cast (which included the finest soprano of the era, Evelyn Herbert), singing soon-to-be Romberg standards such as "Stouthearted Men," "One Kiss," "Softly As In A Morning's Sunrise," "Wanting You," and "Lover, Come Back To Me." It also offered a steadfastly traditional plot. A nobleman turned revolutionary falls in love with a New Orleans shipowner's daughter. But when he is captured and exiled, he is led to believe it was she who betrayed him. She follows him into exile. Reconciliation arrives with news of the French Revolution.

The show delighted Arthur Pollock of the Brooklyn *Eagle* because "it is not musical entertainment of the hard-shelled, tough, brazen kind that so often you cannot avoid seeing in theaters where musical comedy is playing." Similarly, St. John Ervine of the *World* praised it for being "a saxophoneless piece," a "notable . . . return to the stage of light opera." At the New York *Sun* Gilbert W. Gabriel openly confessed to mixed emotions:

> Theoretically, romantic musical comedies [*The New Moon* was called "a romantic musical comedy" in its programs] are all so many brave St. Georges against the dragonfly, Jazz. Everyone of us howls for a return to librettos with big overstuffed plots, choral fireworks, grand ululations by heroes in hand-cuffs and heroines in tears, with overtures, reprises, finaletti and all the other earmarks of old-time operetta scores. Then, as if in answer, along comes something like "The New Moon"—which is certainly and superlatively as good as they come—and we aren't so sure. . . . I feel like a traitor to a noble cause I've championed through these many years of frivoling revues and syncopation contests when I confess that the least romantic elements of "The New Moon" gave me the best time.

Within a few seasons Gabriel was aligned squarely on the side of those dismissing operetta out of hand.

Credit for operetta's luxuriant flourishing in these years belonged almost entirely to two composers: Rudolf Friml and Sigmund Romberg. Because they wrote the operetta triumphs of the twenties and because their careers were more or less parallel chronologically, their public often saw them as two of a kind. Coupling the composers may have begun when both appeared on the Broadway scene almost simultaneously, although Friml's success came overnight while Romberg's was slow and cumulative. Actually, the two men were as dissimilar as they were alike.

Friml followed *The Firefly* with *High Jinks* in 1913 and *Katinka* in 1915, as well as with such lesser hits and failures as *Kitty Darlin'*, *You're in Love*, *Sometime*, and *Glorianna*. For the most part, he wrote in the lush comic opera idiom. He abandoned this style briefly, however, from 1919 to 1923 to write in a lighter musical comedy vein, enjoying considerable success with two now-forgotten works, *The Little Whopper* and *The Blue Kitten*. From the start he rarely offered Broadway more than a single score in any year.

Romberg's earliest efforts were knockabout scores for Shubert revues. His songs were mostly rags or foxtrots, tangos, and other ballroom dance styles, along with the inevitable numbers that allowed bedecked chorus girls to promenade across the stage. The first hint of his real abilities came with his interpolations for *The Blue Paradise* in 1915, but he quickly resumed banging out banalities for the Winter Garden and other Shubert houses. Not until three and a half years (and sixteen shows!) later did *Maytime* call proper attention to his talents. Even then he continued, though with increasing reluctance, to churn out trite ditties for revues and slapdash musical comedies. In 1921 *Blossom Time* added substantially to his reputation, but not until over a decade after his debut, and after contributing part or all of the score for forty shows, did *The Student Prince* firmly establish that reputation in December 1924. Thereafter, apart from *Louie the 14th*, which he had agreed to write before *The Student Prince* opened, and the 1928 hit, *Rosalie*, on which he collaborated with George Gershwin, he confined himself to the style of operetta with which he was most at home. As he himself later joked, he had been a "discontented bachelor" leading an "aimless . . . musical existence" until he "became wedded to Operetta." Significantly, his output dropped immediately as he devoted himself to his real love. Except in 1927, when he offered Broadway four scores, he hardly ever again had more than a single show ready in any one year.

A sharp distinction between the two composers can be seen at the end of their careers. Friml's career ended abruptly after the 1920s. His two operettas that saw the footlights in the early thirties were both discouraging failures. The first, *Luana*, had actually been conceived as a film musical, and was adapted for the stage only after Hollywood dropped it. The second, *Music Hath Charms*, failed to attract audiences even with great prima donna Maria Jeritza as its star. Friml simply could not bend to changing tastes. Perhaps because Romberg was more pliable, he was luckier. His 1930 operetta, *Nina Rosa*, while a failure in New York, was well received abroad. His waltz, "When I Grow Too Old To Dream," written for a 1935 film, *The Night Is Young*, was the most popular song of its day. In 1945 Romberg set to music the sort of bygone Americana Rodgers and Hammerstein made fashionable with *Oklahoma!* The result, *Up in Central Park*, was a smash hit in a season of hits.

Yet although their public careers were similar, Friml and Romberg's private histories, and the artistry those histories shaped, were quite disparate. While both men came from Middle Europe—Friml from Czechoslovakia, Romberg from Hungary—Friml's family was Catholic and poor, Romberg's Jewish and relatively well-to-do.

From childhood on, Friml never considered anything but a life in music. His family went without fuel to purchase an old piano on which the youngster was soon able to pick out melodies he had heard from passing street organs. His skills burgeoned so swiftly that neighbors and relatives contributed funds to send him to the Prague Conservatory, where his talents were instantly recognized, and he was awarded a scholarship. He studied piano with Josef Jiranek and composition with Anton Dvorak. In later years Friml could not remember Dvorak working with any formal plan of instruction. "Dvorak seemed to feel that these theoretical branches were natural with me," he recalled, adding, "I never made any voice progressions that seemed to him incorrect." After graduation, Friml served as accompanist for Jan Kubelik, playing for the violinist throughout Europe and joining him on his 1901 and 1906 American tours. It was on the second tour that Friml decided to make America his home. He spent the next seven years as a concert pianist, composing whenever time permitted.

Romberg began his musical studies on the violin, not the piano. But to his parents, playing a musical instrument was merely a social grace, so as a dutiful son Romberg prepared for a life as a civil engineer. He studied engineering first in a small Hungarian city, then in Vienna. All the while, however, he pursued his musical interests, which included conducting as well as composing. For a brief time he earned extra cash in Vienna by serving as assistant manager at the famous Theater-an-der-Wien, home of so many famous operetta first nights. When he finally advised his parents that he preferred a musical career, they asked him to travel for a year or two before making such an important decision. Romberg spent some time in London, then, in 1909, arrived in the United States. Almost penniless, he took employment in a pencil factory. But he shortly found work as a pianist in a café and was soon conducting his own small orchestra in a fashionable New York restaurant. He then decided not merely to continue in music but to remain in America. The "dancing craze" had begun, and Romberg published three dance numbers which somehow came to the attention of the Shuberts.

Romberg was even-tempered, invariably good-humored. For some reason he never quite mastered the English language, and he soon became celebrated for his malapropisms and his all too literal acceptance of some idioms. Once, in a bridge game, he and Jerome Kern trounced their opponents, and as they left, Romberg informed Kern, "We won our shirts." In contrast, Friml's personality was mercurial, sometimes frighteningly so. His emotions seemed perpetually to be riding a wild roller coaster. One associate, a few years after Friml's death, called him a genius, but suggested that at times he appeared to verge on madness.

When success came to both Romberg and Friml, so did the inevitable interviews. The composers' comments are especially fascinating, revealing not only each man's attitude toward his music but indirectly explaining that music as well. Even the titles of the articles give hints about the nature of their subjects. The famous music magazine of the day, *The Etude*, published an interview with Friml in April 1923, a year and a half before *Rose-Marie* premiered. For the title, the magazine used Friml's own words: "The Mystery of Musical Inspiration." *Theatre Magazine* released Romberg's revelations six years later, just after *The New Moon* had opened. The magazine titled its article "A Peep into the Workshop of a Composer."

Friml's remarks dwelt heavily on passions and instincts. He was proud of his father's "intense fondness" for music, and grateful that his father passed on that love to him. He referred to the piano as his "first inspiration." Friml's devotion to his art was complete:

> To me everything translates into music. Any idea, any poem, any beautiful picture seems to affect my whole being and I am at once conscious of melodies surging up within me. The ocean moves me immensely. I feel its power at once. It is not a question of wanting to compose. I can't help it.

He confessed that even the weather affected his musical moods. "On gloomy days my music is likely to be sad and sentimental. On bright, sparkling, springtime days I want to write music that dances and plays in every measure." Only once in the long article did he mention any of his shows; his minor "comic opera, 'Tumble Inn,' " and then merely to recall that he composed all of it in a thirteen-hour spurt of creativity. He also noted that, while he could and did compose at any hour, "for years I have found that ideas come to me faster and better at two o'clock in the morning than at any other time." Nowhere in his discussion did Friml suggest what he sought in a musical—his reaction to librettos, lyrics, performers, or audiences.

By contrast, the "peep" into Romberg's "workshop" disclosed that the composer thought primarily of mechanics and workaday necessities. He even kept careful count of his labors, calling *The New Moon* his twenty-fifth operetta. (He offered no explanation for his baffling figure.) The gist of his remarks suggested how neatly he compartmentalized every aspect of his craft. To his credit, Romberg looked first at the libretto, a point of indifference to some other composers. "A good book is essential," he began, "It contributes fifty per cent of the success of the operetta."

Regrettably, Romberg made no mention of what he specifically admired in a story, the settings, motifs, or characters that he preferred. Playgoers might be aware of his predilection for bittersweet romantic endings, but readers would garner no hint of this. Instead, Romberg's approach to a plot was not unlike that of earlier composers who had had to create for specific companies. He noted that there must be a prima donna, a tenor, a baritone, and an ingenue. "Then," he continued, not a little ingenuously himself, "the play must be divided into acts, each act marking the division of time. There must be eighteen to twenty-four numbers. No two song numbers of the same kind must follow each other." Moving almost embarrassingly from one obvious point to the next, he then noted that the prima donna and the tenor must each have solos and the two together a duet. Nor could the comedian be neglected. The funny man "must have something funny to sing."

One remarkable revelation was the composer's attitude toward lyrics, an attitude far removed from his more enlightened approach to librettos. "I never read more than the first two or three lines of a lyric," he professed. "After I write the music for what I think the song should be, the lyricist has to write [rewrite?] the words." This curious blind spot led to at least two amusing incidents. Romberg once complimented a collaborator who was anxious to learn the composer's reaction to a new lyric by assuring the lyricist, "It fits." There was nothing sarcastic in Romberg's reply, although the composer may have privately wondered why the lyricist would hope to achieve anything more. Another time, paired again with Jerome Kern in a bridge game, Romberg had to figure out how many of a suit Kern held. Kern obligingly whistled "One Alone." The gambit was unavailing, and Romberg later asked the amazed Kern, "Who knows from lyrics?"

Sadly, Romberg said practically nothing about the nature of the music he composed. Again, he merely compartmentalized—and simplistically at that. He wrote, "Music is in two categories: major and minor. Hungarian, Russian, and Balkan States, Persia and India take music in the minor key; Anglo-Saxon and Latin countries are written in major."

Romberg proudly asserted that he always had his material ready on schedule. He revealed that he composed on an organ and that he wrote in the late afternoon and from midnight to dawn, allying himself with Friml in this respect. "While inspiration must not be pooh-poohed entirely," he concluded, "it is largely a matter of application and perspiration in this field of endeavor, as in every other. . . . Work—that is how I do it." Between shows his attitude was totally at variance with Friml's. At such times, Romberg insisted, he never attempted to compose anything whatsoever.

Yet the sort of librettos for which both Romberg and Friml wrote and the basic musical structures they applied were not all that different. Essentially, they continued the American patterns drawn from older comic opera traditions rather than the more lighthearted, theatrically cynical Viennese-style works. This was especially true of the librettos they selected. Even when, as in two of their reigning hits—*Rose-Marie* and *The Desert Song*—the settings were contemporary (and, in the case of *The Desert Song*, newsworthy), heavy emphasis on exotic local colorings and high romance aligned the works squarely with older traditions. True, *Rose-Marie* was set in 1924 Canada, but the dramatic effect of row upon row of red-coated Mounties and of Indians dressed as totem poles, as well as the alluring grandeur of the Rockies, clothed the story so fancifully that it distracted from it all the while it embellished it.

The Riff wars which furnished the background for *The Desert Song* had been headline news since 1921. In the show, the elusive Red Shadow, actually the son of the French commander, leads the war against his compatriots, triumphs and weds the heroine. Not until shortly before the operetta premiered was the real Riff leader, Abd El Krim, captured and exiled. But like Herbert's *Rose of Algeria*, the mysteries and flavor of a faraway, romanticized land colored and dominated the story. The desert's strange, sometimes menacing, allure stood as background for the more flamboyant French army men and their ladies. Musically, Romberg filled the score with some of the most theatrically exotic chromatics. This was especially evident in certain of the musical passages and in the melodic duet between Middle East and West that began with "One Flower Grows Alone In Your Garden" and was answered by "One Alone," as well as in parts of "Let Love Go" and "The Sabre Song." "The Riff Song," "One Alone," and the title song quickly appealed to a public craving for romantic melody.

Among Friml and Romberg's later successes, the lone works to be set in the United States were *My Maryland* and, in part, *The New Moon*. In *My Maryland*'s retelling of the Barbara Fritchie legend and in *The New Moon*'s totally fictitious adventures, rosy visions of long ago rather than far away provided appropriately picturesque ornamentation. *The Student Prince*, *The Vagabond King*, and *The Three Musketeers* all contrived to remove audiences both in time and space from the mundane problems of their day.

Settings for Romberg and Friml's failed works were much the same. The first act of Friml's *The Wild Rose* was set in Monte Carlo; the second moved to an imaginary principality, Borovina. *Luana* roamed the South Seas. The action of both was laid in the present, as were the beginning and end of *Music Hath Charms*. However, that work, set in and about Venice, flashed back to the eighteenth century for its principal tale.

The only late Friml operetta set in the United States was *The White Eagle*, a musical version of the 1905 hit, *The Squaw Man*. Like *Rose-Marie*, it recounted a romance between a white man and an Indian maiden, this time with an unhappy ending. (This show contained a magnificent Friml piece, "Give Me One Hour." But the rest of the score was lackluster and, like much of the book, dissuadingly derivative.)

Romberg's late twenties and early thirties locales moved with almost Johnsonian expansiveness from Japan (*Cherry Blossoms*) to Peru (*Nina Rosa*), with waystops in Saigon (*East Wind*) and the Ruritanian principality of Zenda (*Princess Flavia*). *The Love Call* and *My Princess* were set in the United States. The former, a musical version of the 1900 hit *Arizona*, unfolded in a not-too-bygone Wild West, while the latter took place in 1927 Greenwich Village. But *My Princess*'s title and the fact that its hero was a prince in disguise revealed its heart was in Graustark.

As a rule Romberg and Friml's love stories proceeded with an almost aggressive earnestness. Comic characters that were introduced rarely had little more than a tenuous connection with the plot and, but for perceptions of what 1920s audiences required, might have been omitted entirely. The heroes of Friml's *The Vagabond King* and *The Three Musketeers*, François Villon and D'Artagnan, evince more humor than most, a trait undoubtedly attributable to the fact that both roles were developed, if not originally conceived, for Dennis King, a performer with acting skills beyond the normal leading man's.

Musically, Romberg and Friml's shows were created along lines that would have pleased Carl Wilmore. Long recitatives, extended choral passages, and prolonged finales of the older comic operas were, for the most part, minimized or eliminated. In this respect there was little difference between these operettas and the better musical comedies of the period. Stories developed largely through spoken dialogue and were interrupted for specific songs when the authors chose to underscore a sentiment or emotion. Waltzes were still employed for the principal love songs and still expected to be high among the shows' hits, but they were not quite the ballroom dances of the Viennese school. One notable exception to the continued utilization of waltzes was the title song of *Rose-Marie*, but, revealingly, Friml expected the show's hit to be his waltz, "The Door Of Her Dreams."

For a careful listener, musical similarities end there. The two men's music echoes their personalities. Although contradictions could be pointed out in either case, Friml's music singes with his often fiery passions; Romberg's warms with his good-natured complacency and sweetness. Compare the driving ardor of "The Song Of The Vagabonds" to the more theatrical impetus of "The Riff Song," or the concentrated intensity of "Love Me Tonight" with the far more languid beauty of "The Desert Song." As far as capturing the essence of the exotic locales they wrote about, neither man was totally flexible or successful. But in this respect, despite Romberg's categorization, the edge appears to fall to Friml. However, the distinction was one of degree rather than kind. Friml's most imaginative evocations, as we have seen, were in *Rose-Marie*. Romberg, for the most part, injected his coloring into his verses rather than his choruses. Note for example the minor chords in the opening of "The Desert Song" or the hint of a pentatonic scale in the start of *Cherry Blossoms*'s "My Own Willow Tree." But the choruses of either song could have been at home in any Romberg work, although he (or his publisher) did add some chords based on fourths to further the Oriental flavor of "My Own Willow Tree." No doubt the ambiance of these shows underscored these tentative colorings and allowed audiences to overlook their shortcomings.

9

The Flowering of Traditional American Operetta: Secondary Figures, Innovators, and Importations

Romberg and Friml dominated the operetta scene of their day, but they did not monopolize it. Victor Herbert's last operetta, *The Dream Girl*, was presented posthumously two weeks before *Rose-Marie* opened, as if to mark a line between two eras. Little was made of the fact that *The Dream Girl* interpolated some discreetly unacknowledged Romberg melodies. It was a modest success. Most American composers failed to enjoy even that small reward. Operettas such as *Tales of Rigo*, *The Enchanted Isle*, *A Noble Rogue*, and *The Well of Romance* disappeared almost as quickly as they arrived. Written largely by talentless, one-shot composers, they served only to harden proliferating anti-operetta sentiments.

But two superior American composers normally associated with musical comedy did surprise their followers with their skill when they tried their hands at the genre. George Gershwin, collaborating with Herbert Stothart, pleaded for a free Russia in *Song of the Flame*. In his title song, Gershwin devised a rallying cry every bit as rousing as Romberg's or Friml's best. Harry Tierney, known primarily as the composer of *Irene*, provided the melodies for *Rio Rita*, an operetta in the tradition of *Naughty Marietta* and *Rose-Marie*. Tierney was able to move effortlessly from his lighter musical comedy tunes to the more lyrical, broad requirements of operetta, offering among others a stirring male chorus number, "The Rangers' Song," the expected love song in 3/4 time, "If You're In Love, You'll Waltz," and the paean to the titular heroine. Yet good as *Rio Rita*'s songs and libretto were, a large measure of the operetta's success was undoubtedly attributable to the particularly loving care and lavish hand Florenz Ziegfeld bestowed on it as the opening attraction of his Ziegfeld Theatre. Whatever contributed to making the operetta so popular an attraction, that a Gershwin or Tierney would compose an operetta implies that attitudes toward the genre were not necessarily hostile.

Relatively few operettas were imported, and of these, fewer still met with cordial receptions. Not a single new Franz Lehar operetta reached Broadway in this period, even though two of his finest works, *Paganini* and *Land of Smiles*, were composed in these years and were popular elsewhere. Leo Fall was represented by a single entry, *Madame Pompadour*, a musical far more welcomed overseas than here. Edward Kunneke's captivating score for *Der Vetter aus Dingsda* was frittered away on an American adaptation by Harry B. Smith, reset in the Civil War period. As *Caroline*, it had a modest run a year and a half before *Rose-Marie* opened. Two more Kunneke works premiered in 1925: *The Love Song*, a hit, if not the tremendous success Hammerstein foresaw, and *Mayflowers*, a failure.

Far and away the most successful Viennese composer in New York's eyes was Emmerich Kalman, the last of the great masters of Viennese operetta's second flowering. He was an incomparable melodist, and only Lehar matched his gift for churning out enduring tunes. But whereas Lehar's music, however beautiful, suggested a certain emotional restraint, an Austrian reserve (despite Lehar's Hungarian origins), Kalman's music never held back. It displayed a Hungarian fervor and emotional openness. A rough parallel can be drawn, musically if not ethnically, between Lehar and Kalman on the one hand and Romberg and Friml on the other.

Yet with the exception of the 1914 hit *Sari* (*Der Zigeunerprimas*), Kalman had suffered nothing but disappointment in America until the mid-twenties. Then his luck here changed abruptly, albeit briefly. His *Countess Maritza*'s 321 performances made it the era's most successful importation, while later in the same 1926–27 season *The Circus Princess* began a profitable six-month stand. Both were quintessential Viennese operettas, set in Hungary and Russia, with nobility and circus folk to heighten their coloring, and filled with sweeping waltzes and ardent gypsy music. *Countess Maritza*'s plot followed the off-again, on-again romance of the countess and the caretaker of her country estate, whom she does not know is an impoverished count. Out of Kalman's memorable score came his most popular song, "Play Gypsies—Dance Gypsies."

Early in the next season Kalman came to America to collaborate with Herbert Stothart and Robert Stolz on the score for a made-to-order operetta, *Golden Dawn*. Arthur Hammerstein had commissioned it as the opening attraction for his new Hammerstein Theatre. Arthur's nephew Oscar worked with Otto Harbach on the libretto. The piece was an oddity. An operetta with distinctly Middle-European music, it was set in Africa and had Americans playing not merely Englishmen and Germans, but, with total seriousness, performing in blackface as Africans. Thanks largely to the work's lovely score and Hammerstein's opulent mounting, it was a success, although by no means the huge success that had been hoped for.

Yet it was not a Viennese composer but a suave, urbane Englishman, Noel Coward, who wrote what might have been the period's biggest imported operetta triumph had not the onset of the Depression deprived it of a run. In writing it in 1929 Coward effectively brought to a close the fifty-year heyday of traditional comic opera and operetta that two other Englishmen, Gilbert and Sullivan, had inaugurated in 1879 with *H.M.S. Pinafore*. Astute and sensitive, he may have even realized he was doing precisely that. At first glance, however, Coward must have seemed the unlikeliest talent to achieve this sort of success. Before *Bitter Sweet* he was known for his sardonic drama, his bright, brittle comedy, and his sophisticated revue material. Yet the excellence evidenced throughout his surprising range of creativity should have alerted playgoers and critics to the possibility that he could also turn his hand masterfully to operetta. Just how masterfully became clear when audiences first heard *Bitter Sweet*'s intelligently constructed story, its superior dialogue and lyrics, and, most of all, its magnificent score—one of the greatest in the history of operetta.

What was not so readily apparent was the way Coward approached his material. *Bitter Sweet*'s story began in 1929 in Lady Shayne's opulent house on Grosvenor Square. The Lady's niece is supposed to marry a man she does not love. She prefers the handsome young leader of a modern jazz orchestra, and she has come to her aunt for advice. The aunt's advice takes the form of reminiscences of her own youth, of her own love for a young musician. Her story flashes back to 1875 when Lady Shayne, then Sarah Millick, elopes with her music teacher, Carl Linden. Sarah changes her name to Sari. Carl takes work in a Viennese café, and the two lovers dream of opening a café of their own. But Carl is killed defending his wife's honor. As Sari Linden, the heroine becomes a great opera star and in the 1890s marries Lord Shayne. At a soiree she agrees to sing some of the old songs she and Carl had so loved. Her singing of them is so poignant that she and her husband realize there will only be one true, sustaining love in her life. The action returns briefly to 1929, and the niece, understanding her aunt's message, goes off to marry the man she most wants.

Flashbacks were not totally new to operetta, nor were the bittersweet sentiments with which Coward imbued the work. But earlier operettas had simply bathed the past in roseate hues. High romance and obvious, if restrained, sentimentality had colored these offerings. What Coward introduced to the genre at this point was a distinct note of nostalgia, both for the bygone days through which the story moved and for operetta itself. But he did not look back through rose-tinted glasses. With his murderous villain and his poignant depiction of Sarah's loneliness, Coward acknowledged the ugliness in past times, and, by indirection, the flaws of the operetta genre.

But he saw the virtues and beauties, too. He brought to his theatrical canvas—his story and his music—a palette awash in sad, lush autumn colors. A wistful plaintiveness runs through much of his score. His best number for the show, "I'll See You Again," was, after all, essentially a farewell, one proclaiming "what has been is past forgetting." Even his bow to operetta's gypsy element eschewed the impassioned ardor of the czardas. In fact, "Zigeuner" went so far as to ignore correct gypsy harmonies and chromatics. But Coward hauntingly evoked a sense of gypsy melancholy through modern modulations, and underscored the feeling in the lyric when Sari confesses, "All I ask of life is just to listen to the songs that you sing." With "Tokay" Coward managed the curious feat of providing a superb choral rouser that was also a not quite traditional drinking song. For Americans at least the show benefited from two compelling assets—the outstanding singing and performing of great English beauty Evelyn Laye and a sumptuous, impeccably tasteful Ziegfeld production.

In his lectures on Coward, Hobart Berolzheimer, retired head of the Philadelphia Library's Theatre Collection, has suggested that, for better or worse, Coward may have initiated the periodic vogues of nostalgia in recent decades. Failing a thorough study of all earlier operettas one cannot say for sure how on target Berolzheimer is. Certainly subtle hints of nostalgia can be detected, as we have seen, as far back as *The Merry Widow*. But as a rule, these nostalgic touches were almost accidental or at least incidental. No previous writers of operetta, it seems, were willing to admit that their works belonged to a tradition that was fading. Obviously Coward recognized this theatrical reality and acted on it. Because of this, *Bitter Sweet* can retain a singular grip on modern audiences.

Yet while traditional operetta was losing ground, Americans were exploring ways to release it from forms they perceived as imprisoning and to revitalize it in wholly native terms. These earliest efforts, however tentative and imaginative, were failures. As far back as 1924, Jerome Kern, always the leader in the musical theatre of his day, attempted to combine musical comedy and operetta by creating a score with longer and more ranging musical lines for a more or less traditional musical comedy libretto. At least initially, that libretto, under the working title *Vanity Fair*, was to have been a highly literate, satirical view of fashionable Park Avenue society. When a smart young lady puts herself up at a charity auction as a maid-for-a-week, a rich playboy whom she has snubbed wins her services and sets about to humiliate her. A story of this sort could, indeed, lend itself to sophisticated, witty comedy. It could just as easily lend itself to slapstick. Unfortunately commercial considerations quickly moved the libretto down several notches, so that by the time the show opened in New York as *Dear Sir* it was something of a stylistic hodgepodge. Furthermore, Kern's score, while venturesome, was not especially memorable. Discouragingly received, the show quickly folded.

In 1926 Arthur Hopkins, one of New York's most enlightened and prestigious producers, offered playgoers what he called a "native opera." *Deep River* had a libretto and lyrics by Lawrence Stallings, a serious playwright best remembered for his co-authorship of *What Price Glory?* The score was by Franke Harling, now a largely forgotten name, who in his prime successfully straddled several musical worlds. A year before *Deep River*'s premiere the Chicago Opera had mounted *A Light from St. Agnes*, Harling's lyric theatricalization of Keats's poem "The Eve of St. Agnes." Later in the twenties the nation whistled his "Beyond The Blue Horizon." Harling's score for *Deep River* took a middle ground without ever compromising its artistic integrity. The songs sustained the story's moods and frequently advanced the action, yet, as with all great operetta material, the best could have been enjoyed away from the show, although sadly, they had no separate currency. In the theatre, played by a thirty-piece orchestra and sung in part by Jules Bledsoe, they must have been magnificent.

Unlike his collaborator, Stallings was uncompromising. Neither subplot nor comic relief distracted from an almost inexorably bitter love story. The central figure is a quadroon named Mugette. She loves Hazard Streatfield, a handsome Kentuckian. But when Hazard kills Mugette's unwanted suitor in a duel, he is forced to flee, and Mugette must resign herself to a life alone. Even the demonic voodoo ceremony, which occupies most of the second act, sustained the sombre moods and advanced the plot development. But in 1926 jazzy musical comedies and older-style operettas remained what Broadway sought. Ruing the loss of such " a lovely thing," the dramatic critic Burns Mantle complained that "the opera-going public would not come down to it, nor could the theatre public rise to it."

A year later, however—in the same 1927–28 season that counted Romberg's *My Maryland* and Friml's *The Three Musketeers* among its hits—the public did rise to the occasion when Kern and Hammerstein found the answer for future operetta in *Show Boat*. Today, *Show Boat* is generally conceded to be the first successful "musical play." Regrettably, this is a meaningless distinction, a distinction rather self-consciously developed in the 1940s to legitimatize the school Rodgers and Hammerstein initiated with *Oklahoma!* and rightfully to acknowledge *Show Boat* as the first outstanding precursor of this school.

Show Boat was actually the first totally American operetta. It was totally American not merely because all its authors were native, or because its setting was American, but because its music leaned heavily on native patterns, its lyrics and dialogue were totally colloquial, and even its unstated philosophic premises discarded fundamentally snobbish European sensibilities and puritanism. By enlarging one of the characters, Joe, and turning him into a sort of one-man tragic chorus, Hammerstein brilliantly tied together an otherwise loosely structured libretto with Joe's touching black fatalism, a fatalism practically unknown in and basically alien to older operetta. *Show Boat* was operetta by virtue of its expansive musical lines and its gorgeously lyrical melodies as well as its highly romantic, sentimental story. That this story moved back in time and into worlds its audiences could view as mildly exotic further linked it with previous decades of comic opera and operetta.

Of course *Show Boat* derived from the large picturesque canvas of Edna Ferber's novel, though Hammerstein necessarily rejected and altered much of Miss Ferber's material when he came to accommodate it to two hours' traffic on a stage. At its heart was a rocky romance between Gaylord Ravenal, a riverboat gambler, and Magnolia, the daughter of a show boat's captain. The marriage is shattered by Ravenal's gambling. After Ravenal leaves, Magnolia takes to singing in night clubs to support herself and her daughter. In the end she elects to return to her family's boat, and years later a burned-out but reformed Ravenal reappears to effect a reunion for the final curtain. If this principal romance had its maudlin moments, a second story, which was played as a minor counterpoint throughout the show, was more maudlin still, depicting the social disintegration of Julie, a riverboat entertainer. Her story introduced the problem of miscegenation, a touchy matter that had been taboo until then on our lyric stage. Thus Hammerstein added heretofore proscribed topics to venturesome philosophic viewpoints in fabricating his text.

Kern may not have been quite as daring as Hammerstein in breaking new ground, but his music perfectly matched the story's turns and fit its moods. As Stanley Green noted, "Its magnificent score is such an essential part of the story that even time-worn sequences still have power to move audiences." From the beginning nothing moved audiences more than "Ol' Man River," in which Kern caught Joe's representative sadness, bitterness, and determination and for which Hammerstein created such an unforgettable lyric. Yet Kern did help introduce new notes into operetta, not only with the blues coloring of "Can't Help Lovin' Dat Man," but by using the more Yankee 4/4 ballad form for the principal love songs—"Make Believe" and "Why Do I Love You?" Operetta's traditional 3/4 time was largely ignored. Aside from the grandiloquent "You Are Love" and a minor song, "Happy The Day," the signature was employed only for scattered incidental passages.

Despite Kern and Hammerstein's towering achievement, the irrefutable evidence of contemporary reviews demonstrates that for many of the era's theatregoers Ziegfeld's largesse and taste were *Show Boat*'s most immediate attractions. His cast, which included Norma Terris, Howard Marsh, Helen Morgan, Edna May Oliver, Charles Winninger, and Jules Bledsoe, was esteemed from the first and with time has become almost legendary. Backing the principals and their supporting players was a chorus of ninety-six, and surrounding the cast were Joseph Urban's eye-filling sets.

When the critics were ready to look beyond Ziegfeld and his contributions, they were quick to realize what a masterpiece *Show Boat* was. They could not, of course, judge its place in theatrical history, nor its durability. But Alexander Woollcott hailed it as "a fine and distinguished achievement," and Robert Garland called it "a wonder and a wow." Stark Young predicted it could lead to popular opera, unable to foresee that in future decades the show would enter some opera repertoires.

That same 1927–28 season also saw an entirely different approach fail, at least temporarily. The Gershwins and George S. Kaufman attempted to create the sort of comic opera that Gilbert and Sullivan might have devised had they been children of the jazz age. *Strike Up the Band* was ultimately to be the first of three works that Ira Gershwin would call "political operetta." In its original form it was an uncompromising satire on war, big business, international politics, and human ego. Its libretto resorted to the same deliciously preposterous inventions that Gilbert had utilized so resourcefully, while its music, although totally of the jazz age, employed the same recitatives, extended concerted passages, and enlarged finales that had been stock-in-trade to Sullivan and his generation. But in the smug heyday of the twenties, the playgoing public would not buy such telling satire, so the show was withdrawn after its Philadelphia tryout.

The year after *Show Boat* Hammerstein and Stallings joined forces with Vincent Youmans to write *Rainbow*. The work was not quite as clearly an operetta as was *Show Boat*. First of all, Hammerstein and Stallings's story was not as sweepingly romantic; it tilted, perhaps, a little more toward its comic side. Set in the 1849 Gold Rush, its central figure is Captain Stanton, who has killed a man in a duel over a woman and is now hiding in disguise as an army parson. Foiling the machinations of the sultry Lotta, he wins the hand of Virginia, his colonel's daughter. For no small number of playgoers these figures were not the evening's cynosure. Many reviews felt the show stealer was Charles Ruggles, in the role of the comic multeer, Nasty. Critics pointed out his resemblance to *Show Boat*'s Cap'n Andy. Also, both shows' heroes had gambling proclivities, and the forlorn Lotta bore a marked likeness to Julie. Though generally underrated, Youmans's score did lack the breadth and grandeur of Kern's material. But by concentrating on another bit of colorful, bygone Americana, Hammerstein and Stallings reaffirmed the direction American operetta was taking.

Rainbow was a quick failure, to some extent because a number of opening night mishaps—a protracted set change, an animal urinating on stage—prejudiced both critics and first-nighters. On the other hand, Romberg's *The New Moon*, which had opened earlier in the same 1928–29 season, became the last book musical for over a decade to run more than five hundred performances. In so doing, it also became the last operetta of the old school to achieve a lengthy run.

In the 1929–30 season, September, October, and November each brought in one important operetta. In no case, however, were these shows stylistically pure, especially since the newer mode had yet to be fully defined. September's entry was Kern and Hammerstein's *Sweet Adeline*; October's, Youmans's *Great Day* (to a text by William Cary Duncan and John Wells); November's *Bitter Sweet*.

Sweet Adeline's story about a young girl's rise to stardom had been employed frequently by musical comedy and by operetta. Both in the unfolding of its plot and in its music, the show hovered precariously between the two genres. Kern's score was slightly constricted by the limitations of his star, Helen Morgan. The show's three most commercial songs—"Why Was I Born?," "Here Am I," and "Don't Ever Leave Me"—were all written for her and appear cut from similar cloth. It was in numbers that did not depend primarily on Miss Morgan that Kern composed in more traditional form and used longer musical lines and more lyrical ranges. These included a polka, "Play Us A Polka Dot"; a folk song, " 'Twas Not So Long Ago"; a march, "Out Of The Blue"; a waltz, "The Sun About To Rise"; and, most of all, a superb men's chorus— with blues harmonies—"A Girl Is On Your Mind." Hammerstein's book was less daring than *Show Boat*'s, but by setting it in a colorful, bygone era and by injecting some genuine pathos into his heroine's life—a pathos developed far more strongly than its tinselly musical comedy counterparts— he embued the text with an earnestness and romantic glow so requisite to operetta. Whatever its shortcomings, the show was well received.

Great Day, on the other hand, was severely panned and quickly withdrawn. Despite Youmans's superior score, nothing seemed to fall in place. Indeed the plot was changed radically during the show's extended tryout. Essentially, it dealt with an impoverished heroine pressed to marry the villain who buys her plantation, and saved only when a young engineer kills the villain. Youmans's hymnlike "Without A Song," the uplifting title number, and the poignant "More Than You Know" survived the debacle.

Under normal conditions both *Sweet Adeline* and November's *Bit-*

ter Sweet would probably have had long, profitable runs. Early in its run *Sweet Adeline* sold out nightly and built what was for the time a sturdy advance sale. But the stock market crash of 1929 changed all that. In the end neither *Sweet Adeline* nor *Bitter Sweet* paid off their investors. Their failures signaled the start of another bleak period for operetta.

10

Operetta in the Depression

The early thirties offered some striking parallels to conditions a half-century earlier when *H.M.S. Pinafore* burst on the theatrical scene. Just as the effects of the 1873 crash were still being felt in 1879, so the country was caught in the morass brought on by the 1929 crash. The Crédit Mobilier and other scandals of the Grant administration and the exposure of the Tweed Ring in New York City found echoes in the Teapot Dome scandal of the Harding administration and, late in 1930, the Seabury investigation of New York Mayor James Walker's corrupt regime. Unemployment was again widespread and growing. Labor unrest was rife, although large-scale strikes similar to the Baltimore and Ohio troubles did not seriously affect the nation until Roosevelt took over and gave labor a free hand. Disenchantment was clearly in the air again, sweeping away the smugness that had characterized the twenties much as it had the immediate post-Civil War years.

Conditions were timely for a revival of a truly comic opera style in the tradition of Gilbert and Sullivan or the early Offenbach opéra bouffes. Of course, the theatre was no longer a place of entertainment for the masses. Radio and sound films had long since drawn away blue-collar and even some white-collar audiences. Still, a large middle-class and upper-class patronage remained loyal to live, legitimate plays. Yet for all their relative affluence, these classes were not untouched or unmoved by the problems of the day. Many were as baffled or disillusioned as the lower classes by the behavior of what a later generation called "the establishment."

The very first musical of the 1930s provided a response to the disillusionment, and belied the argument that operetta was on the wane. On January 14, 1930, a resuscitated *Strike Up the Band* finally reached 42nd Street. This time the "political operetta" was an immediate hit, partly because times had changed and partly because the show had too. Rightly or wrongly, Morrie Ryskind had softened many of Kaufman's sharpest barbs. The whole plot, in fact, was seen as a dream. Angered by a Swiss protest of a tariff that would hinder the sale of Swiss chocolate in America, American chocolate manufcaturer Horace J. Fletcher demands the United States go to war with Switzerland. He even offers to pay for the conflict if it can be known as the Horace J. Fletcher Memorial War. America wins the war when secret Swiss yodeling signals are decoded. By that time, threatened with exposure of the fact that he uses Grade B milk in his chocolates, Fletcher has become an ardent pacifist. But Americans have so enjoyed the war they prepare to engage Russia over a caviar tariff. Gershwin's scintillating score included "Soon," "I've Got A Crush On You," and the resounding title song, as well as something approaching a classic Gilbert and Sullivan patter song, "A Typical Self-Made Man." In this number Bobby Clark's Colonel Holmes (modeled after Colonel House of the Wilson administration) resembled Sir Joseph Porter in *Pinafore*. Holmes tells listeners that he reached the pinnacle of power by, among other things, making mother his ideal. Disillusionment, modification—and marvelous melody—combined to pack houses.

Less than two years later, at the depth of the Depression, the Gershwins, Kaufman, and Ryskind tried their hands again at a similar show. Audiences were even more receptive, and the writers even more artful. The result was *Of Thee I Sing*, a musical that within a few months unashamedly advertised itself as "The Pultizer Prize Operetta." The plot revolved around a presidential election. A political party, unable to develop any genuine issue, has its candidate run on a platform of "Love." The campaign is nearly ruined when the French threaten to go to war with the United States over an alleged breach of promise made by the candidate. The French insist that he had promised to wed the winner of a beauty contest and that the winner was a French girl who is none other than the illegitimate daughter of an illegitimate son of an illegitimate nephew of Napoleon. The tide turns in favor of the candidate when his wife has twins. Obviously, the Supreme Court concludes, a daddy cannot be a "baddy." With the newly elected president unable to marry the French girl, his obligation devolves onto the vice president, whose name no one remembers. To Ira's brilliant and witty lyrics George brought a remarkable array of great melodies. "Love Is Sweeping The Country," "Who Cares?," and the title song all became standards. Besides the songs that could be isolated and sung away from the theatre, Gershwin's score included passages of rhymed recitative and continuity, often as melodic and humorous as the more separable melodies.

Almost another two years later, when the team brought forth a third show, *Let 'Em Eat Cake*, their luck ran out. The authors used the same principals they had in *Of Thee I Sing*, President Wintergreen and Vice-President Throttlebottom. Beset by the Depression's seemingly insoluble programs the politicians turn to fascism, establishing an elite coterie of "blue shirts." But they are deposed by a coup and saved only when sense prevails at the last moment. The musical's failure brought the series of political operettas to an end.

The reason for *Let 'Em Eat Cake*'s quick closing is not hard to see. As a recent summer stock revival demonstrated, it is a suicidally flawed show. Well-plotted, often funny, consistently melodic, it nonetheless ends up uncomfortably depressing, for somewhere along the line Kaufman and Ryskind lost sight of the sunny attitude, the fundamental good nature that this sort of satire demands. *Strike Up the Band* had fun at the expense of big business and militarism; *Of Thee I Sing* spoofed the absurdities of a presidential election, the often inexplicable thinking of the Supreme Court, and, most memorably, the innocuity of the vice presidency. But *Let 'Em Eat Cake* sought humor in fascism, a subject not easily made fun of. Brooks Atkinson lamented that the writers had approached the subject in a "bitter, hysterical mood."

Musically, these shows raise two interesting questions. Since George Gershwin continued to compose in his unequivocal jazz idiom, could he have obliterated the marked separation between musical comedy and operetta that jazz had created? Could he have welded the two into a new, unique genre? The answer to the latter would seem to be a definite "yes." *Of Thee I Sing* and, for all their shortcomings, *Strike Up the Band* and *Let 'Em Eat Cake* are vibrant examples of that very genre. Given the celebrated success of *Of Thee I Sing* (it was able to send out a second company at the nadir of the Depression) and the parallels between the economic situations of 1879 and 1932, the surprise may be that more such modern comic operas were not forthcoming.

To the former question—could Gershwin have obliterated the distinctions between musical comedy and operetta?—the answer is probably not. Gershwin's music is eminently theatrical: at times rousing, at times endearingly sentimental. Indeed, in *American Popular Song* Alec Wilder insists that "Love Is Sweeping The Country" and "Of Thee I Sing" belong "only on the stage," adding, "All of which implies an approach to the realm of what might be called 'swinging operetta.' " "Swinging operetta" is a marvelous term, but it suggests precisely why this new genre could never replace more traditional styles. This new school lacked a certain languorous romanticism that was a hallmark of even the sprightliest old comic operas and operettas. Its music, however effective, misses the glittering touches of the older school at its most mercurial—the broad ranges, the exciting leaps, and the thrilling fioritura passages. These might easily be put down as showmanship, but in proper hands they often conveyed a delight in their own artistry—delight and artistry that could epitomize romance. Additionally, the stories and often the lyrics—admittedly not George's responsibility—were highly cynical in the best modern musical comedy fashion. Of course, here the argument could again be that the writers were reverting to the discarded attitudes of truly comic comic opera.

Several other operettas not only incorporated departures from tradition, but as often as not refused to call themselves operetta. *The Cat and the Fiddle* called itself "a musical love story," *Music in the Air*, "a musical adventure." Jerome Kern's scores for these two gems brilliantly combined traditional Continental material with contemporary American patterns. Had Kern chosen to continue in this vein he might have established a third American variation on traditional operetta, neither the "swinging operetta" of the Gershwins nor the American "musical play" in *Show Boat*'s style, but a unique blend of Middle-European *gemutlichkeit* and Broadway sophistication. Out of these shows came such Kern standards as "The Night Was Made For Love," "She Didn't Say Yes," "The Song Is You," and "I've Told Every Little Star." Less well-known numbers such as *Music in the Air*'s great marching song, "There's A Hill Beyond A Hill," and its inviting waltz, "One More Dance," demonstrate even more clearly the scores' allegiance to traditional operetta fare. As if to emphasize this allegiance Kern kept the music going under much of the action. He had frequent recourse to recitatives and occasional concerted passages.

In retrospect, the librettos Kern worked with for the two shows were not that advanced. Both were set in Europe. Otto Harbach's story for *The Cat and the Fiddle* harked back to the plot Victor Herbert had set to music in *The Only Girl*. The hero and heroine, both composers, are torn asunder by divergent goals. He is a highbrow writer; she, popular. A compromise unites them for a happy ending. Oscar Hammerstein's "adventure," *Music in the Air*, eschewed real villains or dashing heroes, telling instead of good, simple country folk attempting to find success in the big city. In a novel twist, they fail. Apart from the ending, it was a plot employed more often by musical comedy than by operetta, although Hammerstein's taste and art gave his retelling an affectionate quality and dignity earlier versions had lacked.

The Great Waltz, another in a long series of semi- (or pseudo-) biographical operettas, this time of Johann Strauss, and Ralph Benatsky's *White Horse Inn*, whose slim, loose plot (adapted from a Berlin hit) made it almost a revue, relied on spectacular effects afforded by the gigantic stage of Radio City's Center Theatre. These four shows were the only major hits among the operettas of the thirties and very early forties. Of new but traditional operetta there was precious little, and it met with small success. A few shows, such as Benatsky's *Meet My Sister* and Romberg's *Nina Rosa* and *May Wine*, were borderline cases. Out of the nearly one hundred eighty new musicals presented to Broadway in the thirties, fewer than two dozen could be classified as operettas. The percentage in the early forties was still less.

No doubt one reason for the dwindling influx of traditional operetta was the artistic exhaustion of many of its older masters. Friml's two operettas in this period were not memorable; Romberg's six included only one enduring melody ("I Built A Dream One Day"). Nor were the Austrian and German composers more productive. Moreover, by the early thirties the Nazis had taken over in Germany, and even before the Anschluss in 1938 were in favor among the Austrians. Thus, an ugly political coloring soon accentuated operetta's internal weaknesses. Capping the genre's problems was a change in critical attitude. These changes were not directed solely at operetta, but when they were, they took an especially harsh, biased form.

In the propserous, devil-may-care twenties, most critics had taken an easy-going stance when judging musical entertainments. At least one, Alexander Woollcott, had admitted to a double standard—a demanding one for serious drama, a more tolerant one for lyric fripperies. But with the onset of the Depression, virtually all New York newspaper critics adopted tougher standards. Whether this was to justify their jobs and paychecks, or to spare playgoers unnecessary expense cannot be determined. Many a critic seems to have assumed a particularly hard position regarding operetta, possibly because it was somehow seen to be a part of the old discredited order. As soon as the taint of German fascism became attached to the genre, operetta fell into further disfavor with reviewers, although a number of critics did attempt to dissociate politics and art.

When Friml's *Luana*, set rather refreshingly in the South Seas, arrived in September 1930 much of this animus broke out into the open. In the *American* Gabriel confessed, "I've worked up a pretty hopeless operetta complex" and added that he now viewed all of Friml's work as "musical heigh-ho." Writing suavely in the New York *Evening Post*, John Mason Brown bemoaned, "one would have to be something of a confirmed and long starved operetta addict to enjoy this," although he added with gentlemanly objectivity that many in the audience obviously had enjoyed themselves. Robert Littell, critic for the *World*, wrote his entire notice in pidgin English, ending, "Me next boatee to Vienna. Aloha Vienna."

Three nights later, when Romberg's *Nina Rosa* premiered, these same critics took separate tacts. Littell returned to standard English for another snide review, while Gabriel found to his surprise that he enjoyed the evening, commenting, "Even the man who has been groaning about musical romances for the last two or three years admits this is a good one." Brown carefully phrased whatever reservations he held by observing, " 'Nina Rosa' is one of the very best of the elaborate operettas the Messrs. Shubert have produced." But by the time Romberg's next operetta, *East Wind*, opened a year later, the disdain for the genre had exploded once more. "Operetta is, I must confess, a cultivated habit which I cannot pretend to have acquired," Brown wrote, although he added that he had admired *Rio Rita*, *Show Boat*, *Sweet Adeline*, and *The New Moon*. Gabriel characterized the genre as "hokum" and informed his readers that he could not "digest romantic operettas without burping."

By this time Brooks Atkinson had joined the wailing chorus, although his attacks were rarely as shrill as they were to become. Of *East Wind* he observed that it "blew with dark passion" and then jabbed at the composer by concluding, "Mr. Romberg's scores are gratifyingly loud." The very first sentence in Atkinson's review of Friml's *Music Hath Charms* injected political coloring into his distaste for the genre. "Back to the Old Deal in operetta," he moaned.

Nor did a new decade bring a change. When *Night of Love*, with a score by Robert Stolz, opened early in 1941, the headline for Louis Kronenberger's *PM* review read "Nice in 1912." In the New York *Daily News* Burns Mantle regretted the show was "obviously dated," while John Anderson in the *Journal-American* remarked cynically, "the lobby had its advantages." Atkinson, as literate and witty as he was faddish, concluded, "Although everything else in the world is changing with alarming rapidity, Shubert operetta remains standard. 'Night of Love' . . . follows the formula with great skill. There is something almost admirable about the exact proportioning of the tawdriness to the trite."

Yet occasionally some critics did speak out and were still willing to judge each operetta on its own merits. Reviewing the Romberg-Hammerstein work *Sunny River* at the end of the year, Wilella Waldorf began her discussion in the *Post,*

> Max Gordon made a noble gesture last night at the St. James. Mr. Gordon apparently feels that the time has come to bring operetta back to Broadway. A number of theatregoers, including this reviewer, agree with him heartily. We must admit, however, that if "Sunny River" is the best new operetta he could discover, it is not surprising that few producers these days are inclined to give [it] much attention.

And that was probably as honest and fair an assessment of the theatrical picture as the time could offer.

Away from Broadway, at least a few operettas retained their seemingly undying popularity. *Blossom Time* and *The Student Prince* toured incessantly. Hardly a major American theatrical town failed to receive a visit from one or both each season until the late 1940s. The productions were shabby, often utilizing tattered sets and thin-ranked ensembles. At best some fading luminary would be starred or featured as a lure. For several seasons Everett Marshall alternately played Schubert and Prince Karl Franz's tutor, Dr. Engel, in *The Student Prince.* Ann Pennington, the dancing darling of George White's *Scandals,* was brought in for a season or two, looking forlornly like a misplaced Baby Jane and performing to Romberg's more sedate material the same turns she had performed to jazz numbers in the old revues. Supporting players were often embarrassingly inadequate, especially when it came to the full-throated singing that best served these works. At the same time, amateur productions embraced the whole standard repertory.

Hollywood, which in its own way had helped make bygone operettas seem old-fashioned, nonetheless rushed to put many of them before the cameras. At first some were merely cinematic redactions of popular Broadway musicals—*The Vagabond King, The Desert Song, The New Moon, Golden Dawn, Song of the Flame.* These reworkings varied in their fidelity to the originals. *The Vagabond King,* for which

Dennis King recreated his impersonation of Villon, adhered rather closely to the stage version, while *The New Moon* discarded the original libretto and told a different story in a different setting. In no case was the entire stage score retained. Sometimes new melodies, which audiences quickly judged inferior to the original, were inserted. Some old wines, much diluted, were offered in new bottles. Thus, Lehar's *Gypsy Love* was revamped as a vehicle for Lawrence Tibbett and rechristened *The Rogue's Song*, and Oscar Straus's *A Waltz Dream* became *The Smiling Lieutenant*.

This latter film was directed by Ernst Lubitsch and starred Maurice Chevalier. More often than not Lubitsch partnered Chevalier with Jeanette MacDonald (who had also sung opposite Dennis King in Hollywood's *The Vagabond King*). Utilizing both new and derivative material, the trio provided filmgoers with some of the happiest and most successful films of the day. Typical of their made-to-order operettas was their 1929 collaboration *The Love Parade*, a pleasant bit of froth despite the frail songs of Victor Schertzinger and Clifford Grey.

Others also tried their hand at original film operettas. Romberg and Hammerstein wrote *Viennese Nights* and *Children of Dreams*, but neither work enhanced its authors' reputations. Friml, failing to get *Luana* before the cameras, may have been fortunate. In the end, the best loved and best remembered screen operettas were those that coupled Jeanette MacDonald with Nelson Eddy: *Naughty Marietta*, *Maytime*, *The New Moon*, *Sweethearts*, *Rose-Marie*, and *Bitter Sweet*. Produced from 1935 to 1940 they generally played havoc with the librettos and scores on which they were based, but, for the most part, retained a feeling for operetta that Broadway had seemingly lost.

Yet sporadic successes and fitful stirrings suggested there was an audience for romantic, lyrical musicals, which was awaiting the arrival of something fresh and satisfying to rejuvenate the operetta stage. It came soon enough, initially entitled *Away We Go!* And away operetta went, into another renaissance.

11

Oklahoma!: The "Musical Play" or "Folk Operetta"

Although *Oklahoma!*, like *Rose-Marie* before it, was billed as "a musical play," few of the almost universally rapturous notices it received after its March 31, 1943, opening made mention of the fact, or of its implications. Virtually every critic, falling in with the common, comfortable perceptions of the day, branded the show a musical comedy, although some better critics instinctively knew that description didn't truly fit. Burns Mantle of the *Daily News* was typical of the more observant reviewers, praising the work as "the most thoroughly and attractively American musical comedy since Edna Ferber's 'Show Boat' was done by this same Hammerstein and Jerome Kern." In *PM*, Louis Kronenberger welcomed it as "a little more than a musical comedy without being pretentiously so," but recorded that no small part of its merit was its "old-fashioned period quality." Lewis Nichols, who had become the *Times*'s critic when Brooks Atkinson went to spend the war in Russia, was practically alone not only in acknowledging the musical's official billing, but in attempting to go beyond it and pin down its real nature. Nichols ended his notice, "Possibly in addition to being called a musical play, 'Oklahoma' could be called a folk operetta; whatever it is, it is very good."

If, in the first flush of revelation, only Nichols grasped the essence of the show, historians, with time to assimilate it, more clearly appreciated its basic character. By 1950 Cecil Smith could write, admittedly with a certain hesitancy and apparently without acknowledgment of *Show Boat*,

> [Rodgers] made of *Oklahoma!* more of an operetta and less of an out-and-out musical comedy than any of his earlier works. The union of two sympathetic temperaments created the first all-American, non-Broadway musical comedy (or operetta, call it what you will) independent of the manners or traditions of Viennese comic opera or French opera-bouffe . . . *Oklahoma!* turned out to be a people's opera, unpretentious and perfectly modern, but of interest equally to audiences in New York and in Des Moines.

As a result, *Oklahoma!* not only established a Broadway long-run record that stood for many years, but set contemporary marks on the road.

What a felicitous phrase "folk operetta" is! Of course, *Oklahoma!* could not have been "folk operetta" in the classic sense of a folk tale or a folk song. It represented a far cry from such primitive art works. Even the more open modern conception of folk art, in Dan Ben-Amos's words "artistic communication in small groups," could scarcely be applied here. But by introducing more colloquial speech (heretofore more closely associated with musical comedy), by adopting popular musical forms, and by its very subject matter, *Oklahoma!* and its successors consolidated the advances of pioneers such as *Show Boat* and created a new, clearly identifiable type of American operetta. Had the term "folk operetta" caught on, it might have put our whole musical theatre into proper focus, might have eliminated the faddist distinctions and faddist sneering that ensued, and might have allowed Broadway to offer audiences a wider, healthier range of musical shows.

Even "people's opera" would have come close to the mark. Not for nothing is the great Viennese bastion of operetta, dedicated to preserving and reviving past masterpieces, called the *Volksoper*. After all, *Oklahoma!*'s libretto, in its sentimentality, romance, and relative seriousness, was patently more akin to the plots of traditional operetta than to the flippant, often semi-cynical librettos of traditional musical comedy. The determinedly contemporary musicals of the thirties and early forties accentuated the disparity. Rodgers and Hart's *Pal Joey*, for example, depicted the rise and fall of a modern-day nightclub dancer, a calculating little heel who would live off rich, sex-hungry women.

Hammerstein's libretto, however, was set in turn-of-the-century territory that became the state of Oklahoma. Jud, the villain, is not a cynical leech like Joey, but a lonely, embittered, if potentially violent, farmhand. Dreaming above his station he sets his heart on the young lady whose family owns the farm on which he is employed. Out of a certain fear of Jud, the girl, Laurey, agrees to accompany him to a box social, although she really loves the young cowboy, Curly, who also attends the party. Jud's pent-up violence explodes when Laurie rejects his advances. In a fight with Curly that follows Jud is killed. After Curly is tried and acquitted on grounds of justifiable homicide, he and Laurey wed and head off on their honeymoon.

Musically, of course *Oklahoma!* was not as broad as older operettas. There were no extended concerted passages; indeed, there were no concerted passages at all. Nor were there grand finales to each act. The individual songs were softer, smaller-scaled than, say, Herbert's or Friml's. They were far more in line with what Kern had written for operetta from *Show Boat* on. But this was a musical evolution that reflected American and modern tastes as much as anything. It was not a revolutionary departure. The man in the street might not have realized that "Oh, What A Beautiful Mornin'!" was a waltz or cared that Alfred Drake opened the show by beginning the song offstage and without accompaniment. That same man in the street might not have even noticed how cleverly Rodgers captured the horses' clippity-clop in "The Surrey With The Fringe On The Top," or how ingeniously Rodgers suggested the sound of Western square dancing in the uplifting title song. The show's great ballad, "People Will Say We're In Love," was totally traditional in form, with a typical AABA framework. It was the most popular song of the year, and deservedly so. Another waltz, "Out Of My Dreams," unfortunately never attained the popularity that it deserved. More instantly recognizable as a waltz than "Oh, What A Beautiful Mornin'!," its somewhat advanced modulations may have told against it.

Yet although the songs were soft and small-scaled they nonetheless required at least two principals "who were called upon to actually sing the music—*sing*, mind you—not just talk through the lyrics." Hammerstein's old criterion for operetta most certainly applied here. *Oklahoma!*'s musical director, Jay Blackton, apparently agreed. Rejecting the standard musical practice of having the entire chorus sing only songs' melodies, he returned to part singing in the decades-old fashion of traditional operetta.

Even the "folksy" nature of the characters was not totally new to operetta. Itinerant entertainers, laborers, and other lower-class figures had often been heroes and heroines in earlier pieces, although in the earlier works they were as often as not connected by plot contrivances to royalty or, at least, aristocracy. Since Ruritanian principalities had all but disappeared from the globe and surviving kings and queens were largely figureheads, interest in them had naturally evaporated. Once again operetta was merely reflecting a social evolution. (Still, at least one king was to figure prominently in a Rodgers and Hammerstein masterpiece.)

Both Rodgers and Hammerstein, of course, brought to *Oklahoma!* extensive musical and theatrical backgrounds as well as familiarity with the traditional forms of operetta and musical comedy. Richard Rodgers was born in New York City on June 28, 1902, and grew up surrounded by good music. His mother was an accomplished amateur pianist and his father, a doctor, was no mean singer. Rodgers began formal musical studies when he was six. In his early teens he fell in love with musical comedy, and especially with the music of Jerome Kern. At Columbia University he worked on varsity shows, his lyricists there including Lorenz Hart and Oscar Hammerstein II. He and Hart wrote the songs for the 1920 musical, *Poor Little Ritz Girl*, but they did not truly capture Broadway's attention until *The Garrick Gaieties* in 1925. From then until 1942 they provided the songs for many of the most sophisticated, melody-packed musical comedies of their generation.

Oscar Hammerstein II was born into a distinguished theatrical family on July 12, 1895. He entered Columbia University expecting to take a degree in law, but changed his mind when his uncle made him an assistant stage manager. From 1920 to 1940 he worked with nearly a dozen different composers, creating both musical comedies and operettas.

With *Oklahoma!*, Rodgers and Hammerstein's timing was impeccable. If there had been parallels between the early 1930s and the early 1880s, there were parallels between 1943 and 1915. While America had not yet entered the war in 1915, many Americans had already taken sides and had begun to view the Berlin and Vienna homes of operetta with increasing suspicion. The result had been operetta's falling out of favor and the birth of the modern American musical comedy, led by the elegant Princess Theatre shows. By 1943 operetta had long since fallen out of favor again, and its association with Germany and Austria seemingly sealed its coffin. Curiously, by 1943 musical comedy was also languishing. *Pal Joey* and *Lady in the Dark* had recently pointed to new directions, but Broadway seemed disinclined to follow up on their trail-blazing. Something refreshingly new was called for.

The war had brought to the fore an interest in Americana. Folk song, long the province of a small band of scholars and the "folk" who sang it, was starting to enjoy a wider vogue. Of course, Kern had miraculously combined accepted theatrical musical forms with American idioms in his music for *Show Boat*, while Hammerstein's honest approach to the riverboat performers, the riverboat gamblers, and the stevedores of his tale was a significant breakaway. But in the prosperous, hedonistic twenties, *Show Boat*, for all the praise heaped on it, was little more than a tentative exploration. Now Rodgers and Hammerstein moved this totally American operetta into the mainstream.

That they were aiming for something refreshingly new cannot be gainsaid, yet the question of how new remains. Their original title, *Away We Go!*, had something frivolous about it, something aligning it, perhaps, more with musical comedies of the day than with traditional operetta. And possibly something could be made of the show's opening number being a waltz—but not too much. One has the feeling that intuitively if not expressly Rodgers and Hammerstein understood that in its own way *Oklahoma!*, again like *Rose-Marie* before it, was essentially an "operetta—[a] musical play with music and plot welded together in skillful cohesion." They and their associates, who agreed on *Oklahoma!*'s billing, could not have been unaware that "musical play" had been employed as a synonym or euphemism for operetta since the turn of the century and especially in the twenties. Yet whatever Rodgers and Hammerstein intended, thanks to *Oklahoma!* folk-operetta was here to stay and was to become the reigning mode for nearly a quarter-century.

Nevertheless, for all the mystique these new "musical plays" soon developed, they did not take over at once, nor did attacks on older genres at first suggest that modern musical plays were the be-all and end-all of our musical stages. Indeed, older-style musical comedy flourished during the succeeding season (along with the brilliant, innovative *One Touch of Venus*), and even revivals of old operettas enjoyed largely favorable receptions. *The Vagabond King* and *The Student Prince* both received cordial notices. Howard Barnes of the *Herald Tribune* may have seen the Friml piece as "creaking badly at the joints," but both works could obviously call on a vast reservoir of affection. Kronenberger of *PM*, though he regretted that the Shuberts had hustled in a tacky, somewhat inadequate road-company production of Romberg's masterpiece, still wrote, "If you liked it once, you'll like it again."

A far greater success was a revival of nothing less than *The Merry Widow*. Several critics complained about the libretto's stale comedy, but they were happy to look beyond this minor failing and let themselves be beguiled. Kronenberger went so far as to rejoice that "last night's production, like *Oklahoma!*, gives the town something to be proud and happy over." Lest potential playgoers stay away because Lehar had "played ball with the Nazis," Kronenberger assured his readers the composer was not receiving any royalties. Nor was Kronenberger alone in comparing the excellences of *The Merry Widow* to *Oklahoma!* Since by chance the two operettas were playing directly across the street from one another, John Chapman of the *Daily News*, equally delighted with the pair, urged his readers to hasten to see both.

His reasons were as much didactic as they were to afford his followers entertainment, for he noted, "West Forty-Fourth St. now embraces a capsule history of the musical comedy business."

In the next season, an operetta squarely in *Blossom Time*'s tradition, *Song of Norway*, was welcomed vigorously and became a smash hit. The show used Grieg themes to embellish a fictitious account of the composer's life. Whatever its weaknesses, the show had in its corner Edwin Lester's sumptuous production and the finest singing Broadway had heard in many seasons. (that operetta music had changed in some ways was brought home by the failure of *Song of Norway*'s great singers—Lawrence Brooks, Helena Bliss, and Robert Shafer—to carve out important careers for themselves on the strength of their fine performances.)

A second operetta, *Rhapsody*, with music by Fritz Kreisler, who had composed the successful *Apple Blossoms* a quarter-century before, folded quickly after scathing notices, but, significantly, the show was attacked for being bad operetta, not simply for being an operetta. Thus, Nichols wrote in the *Times*, "The songs are there and one day someone may write an operetta around them: 'Rhapsody' definitely is not it." Kronenberger assailed the show as " a travesty not only of court life but of grand-scale operetta as well."

Similarly, when *Marinka*, composed by Emmerich Kalman, opened in the summer of 1945 those critics who failed to enjoy it again attacked it not as operetta but as a poor example of the genre. Kronenberger suggested,

> An operetta version of the *Mayerling* story, with the right lilting music, the right *Alt Wien* atmosphere, the right pretty girls and romantic lovers, would seem made to order for July. But by and large *Marinka* has let its chances slide . . . all the songs could have come out of any operetta written since McKinley was in the White House and an orchestrational attempt to liven their beat only detracts in the end from their schmalziness.

Obviously, Kronenberger's feelings were slightly ambivalent, as implied by his injection of the term "schmalziness." Even his remark "made to order for July" categorizes such entertainment as the sort of frivolity critics used to call "a summer show." But his suggestion that an operetta could still be written and his displeasure with orchestrations that destroyed the music's true flavor indicate no little surviving respect.

It fell to Brooks Atkinson, on his return from Russia after the war, to release an unrelenting fusillade against operetta in the pages of the *Times*. In September 1946 two operettas reached Broadway, and Atkinson opened fire. First to arrive was Lehar's long-delayed *The Land of Smiles*, with its original star, the great Viennese tenor Richard Tauber, re-creating the role written for him—that of Sou-Chong, the Chinese prince who nobly releases a Western girl from her vows when he realizes East and West are irreconcilable. Tauber's great song was the aria always identified with him, "Dein ist mein ganzes Herz" ("Yours Is My Heart Alone"). The show divided the critics, many of whom recognized the problems Tauber had suffered as a Jew and a refugee, the affection in which he was held, and the fact that *Das Land des Lachelns* had been a worldwide success. Vernon Rice observed in the *Post*, "It's become an accepted fact that most operettas have bad books," and Kronenberger dismissed it as "very dreary, very moldy operetta." Yet they and their colleagues still could see the show's good points and were willing to await a better operetta. Atkinson, however, although he enjoyed some of the music, sounded a slightly shrill note, condemning "antimacassar operetta" and complaining, "Why not treat the score as good concert music and spare the theatre?"

When *Gypsy Lady* opened, most critics were displeased with the claptrap book, which had been devised to support an assemblage of early Victor Herbert songs. *PM*'s headline read, "Victor Herbert Goes It Alone," while Ward Morehouse of the *Sun* was saddened that "The delightful music in 'Gypsy Lady' is not enough to offset the gawky book." Once again, Atkinson used his review to attack operetta in general, denigrating one of the great American masterpieces in the process. As far as he was concerned, "The odor of 'The Student Prince' hangs round it." He clearly meant an off-odor.

When another *Blossom Time*-like pseudo-biographical operetta appeared a year later, Atkinson continued his onslaught. *Music in My Heart*, he wrote, was "an evening of old hat . . . which hurried out of fashion with 'Of Thee I Sing' and 'Brigadoon.' " In his annoyance the critic overlooked the success of the similar, if superior, *Song of Norway* just a few seasons back, as well as the fact that both *Of Thee I Sing* and *Brigadoon* were essentially operettas, although admittedly of newer schools. Other critics again were more tolerant. For example, Robert Coleman of the *Mirror*, while granting that the show "gave operetta another black eye," was not ready to see the whole genre down for the count.

Even in a show that was not an obviously traditional operetta, Atkinson could find occasion for a gratuitous stab. Discussing Heitor Villa-Lobos's *Magdalena*, Atkinson insisted he felt "as though nothing had been accomplished since 'The Prince of Pilsen.' " He suggested Villa-Lobos and his comrades look to *Brigadoon*, *Finian's Rainbow*, *Street Scene*, and *Oklahoma!* for models.

But it remained for an October 1948 entry to prompt Atkinson to fire all guns full blast. That show was Sigmund Romberg's lyric version of Edward Sheldon's play *Romance*, retitled *My Romance*. Its story was the same as that in which Doris Keene had played so successfully from 1913 on, the foredoomed romance of two social unequals: a minister and an opera singer. Anne Jeffreys and Lawrence Brooks (who had portrayed Grieg in *Song of Norway*) played the parts ably and sang them radiantly. Romberg's melodies, charming period pieces, were memorable. But all this left Atkinson unsatisfied, and not unmoved, for *My Romance* was an unabashed operetta. The critic again used his review space to launch a violent attack on operetta itself. His opening paragraph read:

> Lovers of the dramatic unities should study "My Romance," which was put on at the Shubert last evening. It is standard operetta with standard routines and situations that have not changed through the years. At one time the authors and composers may have believed in those ritualized gestures toward stock romance. But they are pure formula now with high society, passionate love that tears the vocal chords apart and sets the brasses and drums to roaring in the orchestra pit, a theme song of mobility, elegance and boredom.

In between this and his conclusion that the show was "pretentious fiddle-faddle," Atkinson maintained a concerted barrage.

Atkinson's attack is revealing for a number of reasons. He almost certainly decided to use his great prestige to flail out of existence what he deemed a worthless style. To this end he loaded his criticism. The term "standard operetta" crops up five times in seven paragraphs, including the observation that "Mr. Romberg's musical routines . . . are as old and usually as uninspired as standard operetta." Words such as "stock," "routine," and "commonplace" abound. Of course, the critic's insistence on the adjective "standard" makes one wonder what, if anything, he considered non-standard operetta. Did he recognize that many of the newer musical plays he so delighted in were operettas, but decide not to use that term in the interest of propagandizing for the modern school? On top of this, Atkinson injected a political coloring by referring to "high society" and "mobility" (as if that were a capitalistic fantasy), and by announcing he had left the show early "in the interests of progressive morning journalism." There was even a hint of native chauvinism when he accused Romberg of running "through a whole library of formal European mannerisms."

After Atkinson's retirement, his successor, Howard Taubman, continued the assualt, raising the uneasy, if unlikely, possibility that through some byzantine reasoning an anti-operetta stance had become *Times* policy. A harmless spoof of the older schools, Rick Besoyan's *The Student Gypsy*, or *The Prince of Liederkranz*, served as a launching pad for Taubman's biggest salvo. Like Atkinson's review of *My Romance*, paragraph upon paragraph of Taubman's notice was peppered with gratuitous volleys: "Do you think someone ought to whisper to him [Besoyan] that the silly, extravagant operettas of rapid [vapid?] romances in gingerbread kingdoms have been extinct for, lo, these many years?"/"The empty operettas were demolished long ago"/"the ludicrous stuff that was offered to the public with a straight face in the operettas of unlamented memory." Ignoring the failure of any number of 1920s revue sketches that parodied operetta, in order to dissuade playgoers from flocking to the decade's great operetta hits, Taubman concluded:

> Had Mr. Besoyan been around 50 or 60 years ago, he would have shortened the careers of the "Blossom Times" and "Student Princes." Now he must face the troubling question whether there is anything he

> or anyone else can do that's funnier than the original would be if revived straight. Anyone for "The Student Prince"?

As recently as the late seventies, *Times*'s music critic Harold Schonberg, commenting on a City Opera revival of *Naughty Marietta*, lambasted Herbert for composing Muzak rather than music.

Of course, in his day Atkinson was not totally alone in his sentiments. But his respected position and the consistency of his attacks gave his view exceptional exposure and potency. It was undoubtedly coincidental that *My Romance* was the last traditional operetta offered by either Romberg or his producers, the Messrs. Shubert—both of whom had led the way in promoting operettas in earlier days. (A Romberg score was heard after his death, but it was in a much lighter vein.) Still, it was not nearly so coincidental that nothing approaching traditional operetta was offered on Broadway for a number of seasons after *My Romance* folded. Not until 1953 did another brave Broadway, and that one, *Kismet*, stands out not only because of its rousing success but also simply because it was so unique.

Kismet was pure hokum, tuneful, colorful, and often surprisingly literate, but nonetheless hokum, not meant to be taken as an earnest statement or genuine reflection of the world around it. Derived from Edward Knoblock's old warhorse, so long identified with Otis Skinner, it was set to melodies developed from themes by Alexsandr Borodin, one of which, based on a theme from "The Dance Of The Polovtsian Maidens" from *Prince Igor* achieved great popularity as "A Stranger In Paradise." The Arabian Nights story featured the maneuvers of the wily Hajj, who becomes emir for a day and marries his daughter to a caliph. Lavishly mounted (appropriately at the Ziegfeld Theatre), the show offered the great singing and incomparably bravura acting of Alfred Drake as additional enticement for playgoers.

All this while, however, "non-standard" or "folk" operetta or "musical play" had a field day, a heyday as many would assuredly say. Nevertheless musical comedies continued to far outnumber operettas. But there can be no question that the conscious artfulness advocated by proponents and creators of the new musical plays spilled over into these lighter entertainments, often imbuing them with an imaginative and pervasive sense of style that had been lacking in earlier musical comedies. Inventive, often fully realized musical comedies such as *One Touch of Venus; Love Life; Kiss Me, Kate; Guys and Dolls;* and *Gypsy* regularly gave operettas a run for their money, although until the great wave of musical plays ran its course, the best new operettas outlasted their musical comedy competition. Thus, of the two biggest hits in the 1948–49 season, *South Pacific* far outran *Kiss Me, Kate;* while of the two biggest successes of the 1950–51 season, *The King and I* and *Guys and Dolls*, the operetta edged out the musical comedy.

But if what were perceived as the virtues of the new musical plays often had beneficial effects on competing musical comedies, their effects on other musical plays were more pervasive. Inevitably a lot of theorizing and propagandizing followed *Oklahoma!*'s phenomenal success. Some of those theorists and propagandists had vested interests in promoting the new form; others were essentially opportunistic faddists, anxious to jump on the byline bandwagon. One of the most honest and modest assessments came from the presiding genius of the new school, Richard Rodgers. Looking back in his autobiography he mused:

> *Oklahoma!* did, of course, have an effect on the musicals that came after it. Everyone suddenly became "integration"-conscious, as if the idea of welding together song, story and dance had never been thought of before. There were also a number of costume musicals, and no self-respecting production dared open without at least one "serious" ballet.
>
> But in a broader sense I feel that the chief influence of *Oklahoma!* was simply to serve notice that when writers came up with something different, and if it had merit, there would be a large and receptive audience waiting for it. Librettists, lyricists and composers now had a new incentive to explore a multitude of themes and techniques within the framework of the commercial musical theatre. From *Oklahoma!* on, with only rare exceptions, the memorable productions have been those daring to break free of the conventional mold. Freedom is the sunniest climate for creativity, and *Oklahoma!* certainly contributed to that climate.

Apart from the questionable statement that memorable modern productions have been those daring to break free of the conventional mold, Rodgers's analysis offers the best starting point for an examination of the musical renaissance that *Oklahoma!* spurred. In fact, some of the most memorable productions were to no small extent memorable simply because they observed the new conventions Rodgers and Hammerstein helped to mold. Let's look at these conventions and the developments and musicals that followed.

Probably nothing was made as much of by commentators in the forties and afterward as the "integration" of lyrics, music, and drama in *Oklahoma!* Rodgers displayed his own integrity and sense of theatrical history by minimizing this claim. Of course, many earlier musicals were shamefully slapdash, bringing in characters merely for a funny scene or introducing a song that had no bearing on the plot. But in this, musical comedies were by far the chief offenders. As recently as nine years before *Oklahoma!*, *Anything Goes* had switched plots in mid-rehearsal. For the most part, comic operas and operettas from Gilbert and Sullivan on had skillfully wedded words and music, allowing songs to develop characterization, to amplify mood, and to further plot. Those words may not always have been the most imaginative or poetic. Rare was a lyricist with Gilbert's succinct, biting wit or Hammerstein's fresh, homey metaphors. Indeed the banality of so many operetta lyrics had long been recognized as one of the form's major failings. Richard Watts, Jr.'s parting shot in his review of *The Desert Song* so hit the mark that it has often been quoted: "The lyrics gave every indication that W. S. Gilbert lived and died in vain." And if the truth be told, *The Desert Song*'s lyrics were better than average, written as they were in part by none other than Oscar Hammerstein! But if the quality of so many operettas' lyrics (as well as their dialogue) was discouraging, their aptness to their story was almost always evident.

From a textual point of view, then, *Oklahoma!* marked no electrifying advance in the integration of words and music. But Rodgers carefully talked of three conceptions—"song, story and dance"—adding that after *Oklahoma!* "serious ballet" became de rigueur. Here, of course, was the nub of a major change that *Oklahoma!* did bring about, although, again, it was not quite the first show to do so. Ballet had been featured importantly in some earlier Rodgers and Hart musical comedies, and in the case of *On Your Toes*'s "Slaughter On Tenth Avenue" could be said to have had a tentative role in advancing the plot. Similarly, the celebrated "Totem Tom Tom" dance in *Rose-Marie*, while largely ornamental, nevertheless had a slim connection with the story. But Agnes de Mille's "Dream Ballet" in *Oklahoma!*, by detailing the frightening consequences the heroine envisages and thereby causing her to decide to accompany the villain to a box social, did integrate the dance into the fabric of the musical far more than had ever been done before. The small hippity-hop routines that had sufficed operetta until then and the long line of tapping choruses that had served musical comedy so thoroughly were instantly considered outmoded.

Ballet choreographers were suddenly in the spotlight, and not just ballet but "dream ballets" threatened to become a plague. The very next season even musical comedies rushed to fill the demand. Agnes de Mille was able to provide an acceptable dream ballet for *One Touch of Venus*, but George Balanchine's dream ballet for an Arabian Nights absurdity called *Dream with Music* was panned along with the show. For a second Arabian Nights absurdity, *Allah Be Praised*, Jack Cole created ballet out of a slow-motion baseball game. Every major hit of the 1944–45 season contained at least one major ballet. Some, such as Helen Tamaris's evocative "Currier and Ives Ballet" for *Up in Central Park* (which presented an exquisite stage picture of the age in which the action was set and was the high point of the show), had relatively little to do with the plot. Some, such as Jerome Robbins's dynamic material for the musical comedy *On the Town*, helped underscore the tenor of the whole show and, though it could not be foreseen at the time, pointed to a direction for dance on popular musical stages. When Rodgers and Hammerstein brought in their second collaboration, *Carousel*, at the end of the season, they opened the show with a ballet (or, perhaps more correctly, a pantomime), "The Carousel Waltz," which adroitly took care of some basic exposition. But significantly, this was their last show, excepting *The King and I*, in which ballet would figure prominently. As early as 1945, Wilella Waldorf complained in the *Post*, "The 'Oklahoma' formula is becoming a bit monotonous, and so are Miss De Mille's ballets." Astute showmen that they were, Rodgers and Hammerstein sensed a surfeit might be setting in. Nor did they want to slavishly follow their own examples.

After the 1944–45 season, ballet was used more selectively in musical plays, and generally with great intelligence. In a few later instances, notably *West Side Story*, it seemed almost the reason for the show. When it was employed carelessly, as in the otherwise superior *A Tree Grows in Brooklyn*, it bogged down the entertainment. Thus the respectability and vogue *Oklahoma!* created for ballet on Broadway allowed America's great choreographers finally to enjoy widespread commercial success. If many of the later great operettas had little in the way of major ballets—a reversion to the minimal dancing in older comic operas—the dancing in many Broadway musicals thereafter possessed an artistry and dramatic intensity never before realized on the popular stage.

"There were also a number of costume musicals," Rodgers continued. He was, of course, correct, although his expression "costume musical" leaves him open to misconception. All musicals—all Broadway shows for that matter—are costumed. But Rodgers was obviously speaking of period costumes, and here he gingerly touched on one of the most important points of the new folk operetta. Most of the early examples, as well as many of the better musical comedies that borrowed much from them, were period pieces. This trip to the past made, in its own way, a statement about the romantic nature of these newer operettas every bit as telling as the geographically remote and quaint settings of older operettas.

As mentioned before, a safe distinction between operetta and musical comedy had been that operetta had more frequently than not been set in Europe or in some even more fanciful or exotic place and, with the notable exception of the waltz-operettas of *The Merry Widow* era, had often been removed in time, while most musical comedies had been steadfastly contemporary and domestic in their settings. With the coming of jazz musical comedies and especially in the thirties, these distinctions became strikingly apparent—the few exceptions scarcely testing the unwritten rule. But after *Oklahoma!*, yesterdays, especially turn-of-the-century days, helped to set the requisite romantic mood. By exploiting the same practice, some new musical comedies blurred distinctions. A list including *Bloomer Girl, Up in Central Park, Carousel, St. Louis Woman, Annie Get Your Gun, High Button Shoes*, and *Miss Liberty* demonstrates how effective and all-encompassing was the use of period.

Only two shows in this list, *Up in Central Park* and *Carousel*, were incontestably operettas. Oddly enough, the former, by that past master of soaring romantic material Sigmund Romberg, was musically less lyrical and expansive. No doubt Romberg was clearly and sensibly responding not merely to changing conceptions of theatre music but to the relatively lightweight libretto Herbert and Dorothy Fields had supplied him. By playing out their love story against the background of the Tweed Ring scandal—their hero was a muckraking reporter; their heroine the daughter of a Tweed crony—the Fieldses sought to present both another beguiling glimpse of period Americana and a passing bit of righteous social protest. But their treatment of the material was more entertaining than artful. Lightening his musical material accordingly, Romberg nevertheless tried to retain certain operetta traditions while creating a wholly modern score. The results were not totally satisfying. In keeping with contemporary practice he framed his major song, "Close As Pages In A Book," as a ballad, but wrote for it an overblown melody. He also provided another rousing chorus number, but one that, instead of celebrating Mounties or Texas Rangers or Canucks, was transformed by Miss Fields into another local salute not unlike Hammerstein's title song for *Oklahoma!* "The Big Back Yard" sang Central Park's praises. Sweet, gossamer "April Snow" was not unlike *Maytime*'s fragile "In Our Little Home Sweet Home."

Carousel, probably Rodgers and Hammerstein's master opus, told a more believable, tragic tale that not even elements of fantasy at the end could impugn. Hammerstein reset Ferenc Molnar's *Liliom* in a nineteenth-century New England fishing village. There the brash carnival barker, Billy Bigelow, falls in love with demure Julie Jordan. They wed, and when Billy learns that Julie is expecting a baby, he plans a robbery to obtain the money they will need. The robbery misfires, and Billy kills himself rather than be caught. In heaven he is granted a chance to return to earth to see how his wife and child have fared. But he becomes unnerved, slaps his daughter, and returns to heaven unfulfilled. Julie and her daughter are left to face life alone.

Hammerstein treated the story with such understanding and compassion that its most farfetched moments of fantasy were not merely acceptable but deeply affecting. Rodgers's great score included the play's beautiful symphonic opening, "The Carousel Waltz," an extended soliloquy that vividly caught the troubled hero's moods, and the powerful, yet carefully controlled ballad, "If I Loved You." The occasional passages of recitative, as in the lead-in to the same "If I Loved You," had a singular lyricism, a far cry from the flatness of most traditional recitative. However, for much of the show the writers brilliantly incorporated into the verses and even the choruses of their songs what less imaginative hands would have treated as exposition or mere dialogue. Witness, for example, the verse Carrie sings to "Mister Snow." The proportion of music to dialogue was tilted noticeably in favor of music.

With *Allegro*, late in the forties, Rodgers and Hammerstein again led the way, this time by bringing settings back into the present day, but *The Golden Apple, My Fair Lady, Candide, Saratoga, Greenwillow, Camelot*, and *Fiddler on the Roof* continued to attest to the past's appeal. Initially this past was almost exclusively the American past. And it was here that Rodgers and Hammerstein (and, of course, *Show Boat*) set a doubly unique stamp on mid- twentieth-century operetta, for not only were these shows set in America—with all that implies, such as American musical idiom and American philosophic undercurrents—but they turned to relatively ordinary people instead of aristocrats for their subjects. Just as *Show Boat* had dealt with riverboat people, so these later shows dealt with cowboys, fishermen, and carnival people, newsmen, city politicians, even ghetto gangs—the "folk" and their middle-class associates. Thus not merely America but the very sort of Americans depicted determined the tenor of these operettas. Something of a "folk" point of view necessarily imbued these stories with Americans' fundamental optimism, their curious dichotomy of conservatism and protest. Not that optimism, conservatism, or protest had been wanting in older schools. The facile optimism of those shows' generally happy endings, their occasional overthrowing of "the establishment" as represented by villainous rulers or lecherous generals, coupled with their fundamental acceptance of the status quo meant that these attitudes were not unknown.

Now these same attitudes were often brought to the fore. There was a singularly American exuberance in cowhands whooping that they know they belong to the land and the land they belong to is grand, or in a nurse confessing she cannot dismiss her optimism, however cockeyed. Similarly, protest found conspicuous vent in a sailor's bitter realization that racial prejudice has to be carefully taught, in affluent professionals despairing that they must move at the same tempo morning and night, allegro.

But two other elements rarer in older operetta were also present, although not so immediately apparent. The first would seem to be distinctly un-American—an absence of that very mobility that Americans have always prided themselves on and that Atkinson had scoffingly included as one of the clichés of earlier works. In many an older operetta, a sweet, impoverished flower girl or a winsome waif would win the hand of a prince or rich man by the final curtain. But *Oklahoma!*'s hero and heroine head off on their honeymoon expecting to return to farm their land; *Carousel*'s heroine can look forward to widowhood and probably no noticeable improvement in her way of life; *Allegro*'s hero leaves the big city for the humdrum simplicities of a small town; and *West Side Story*'s heroine must try to pick up the pieces after her lover is stabbed to death. Life, these shows seemed to suggest, was often making the best of an unhappy compromise. And this element of fatalism—which Hammerstein had first developed so beautifully in *Show Boat*—coming hard upon apparent immobility, gave many of these works a unique stamp. An occasional older operetta had much the same outlook—witness again Lehar's *Gypsy Love* or possibly even *The Student Prince*—but the outlook was exceptional. As a result of this more or less persistent undercurrent, newer musical plays, however artful and enthralling, often sent their audiences home in a somber mood. For the joyous fripperies of past theatrical evenings playgoers had to look elsewhere, or simply look back longingly.

Of course it was not merely fatalism that injected a somber or at least discomforting note into the musicals. Implicitly or explicitly, shows such as *Allegro* and *West Side Story* were propelled in no small measure by insistent social protest, or at least social commentary. If *Allegro* followed Joseph Taylor, Jr., from birth through school, courtship and marriage, internship, rising professional status, and finally rebellion, it also recorded his family's political fickleness. In the late twenties Hoover is praised; by the early thirties he is damned and Roosevelt reluctantly accepted; with the return of prosperity Roosevelt is condemned as a traitor to his class and Hoover fondly remembered. Songs as well as dialogue kept up the jabs. In a song called "Money Isn't Everything" a group of poor people sneeringly spout rich men's platitudes about how little money can really buy—except, as they bitterly add, security and luxury. In another song polite cocktail party conversations are dismissed as "Phoney baloney, tripe and trash!" or, as the song title continues, "Ya-Ta-Ta." Throughout the show, the tired businessman and his wife who comprised so large a portion of the audience found their world and their very sentiments sniped at.

If some discomforting underlying attitudes shaped the new school and remained basically unchanged for a quarter-century or longer, surface details of plot and setting did change. When Rodgers and Hammerstein had let *Allegro*'s tale move from its opening at the turn of the century to a contemporary conclusion, they had pointed to one direction this change would take. (In the very next season an imaginative, if flawed, musical comedy called *Love Life* moved its principals in semi-fantasy from colonial times to the present.)

With *South Pacific* Rodgers and Hammerstein took another step away, setting their action on lush, exotic islands and making their hero a Frenchman. The two love stories interwoven in the libretto both dealt with a basic comic opera motif—a social disparity between the lovers. The cultured French hero loves a hickish American army nurse, and the white American lieutenant loves a native girl. Although Hammerstein injected a strong attack on racial prejudice in the song "You've Got To Be Carefully Taught" (the show encountered problems in the South because of it) and although at one point the heroine balks at pursuing her romance with the Frenchman because he has had a Polynesian mistress, Hammerstein skirted the real issue, dramatically at least, by killing off the lieutenant before he could wed the native girl. Even the death of this secondary hero added a depressing note. Yet the story as a whole, and Rodgers's vaulting score were so ardently romantic that in the end they set the tone for the show, and alone were remembered. Fortunate in having Ezio Pinza's great operatic basso and William Tabbert's fine theatre voice, Rodgers wrote unrestrictedly in songs such as "Some Enchanted Evening," "Younger Than Springtime," and the superb, dark waltz, "This Nearly Was Mine." For the popular Mary Martin, Rodgers composed in a somewhat lighter vein.

By 1950, then, modern operetta no longer felt compelled to resort to America's past. It had regained its freedom to move anywhere or to any time. In 1951 Rodgers and Hammerstein abandoned virtually all connection with America and even brought royalty back to the operetta stage when they moved to nineteenth-century Siam for *The King and I*. (Their one small bow to America was a ballet purporting to be a Siamese version of *Uncle Tom's Cabin.*) Based on actual history as recounted in the book *Anna and the King of Siam*, the principal plot related the essentially platonic love affair between two disparate figures—an English schoolmarm and the oriental potentate whose children she is brought in to teach. A subplot told the tragic love tale of two young courtiers. Hammerstein again used his story to preach tolerance and compassion. For his stars, Gertrude Lawrence and Yul Brynner, Rodgers composed embraceable, sometimes sweeping but essentially restricted songs such as "Hello, Young Lovers," "I Whistle A Happy Tune," "Getting To Know You," and "Shall We Dance?" He was able to let himself go musically when writing for his better-voiced secondary players, awarding them such piercing numbers as "We Kiss In A Shadow," "I Have Dreamed," and "Something Wonderful." The ballet music and some incidental passages also allowed Rodgers to demonstrate the gamut of his musical invention. The warmth of his music, the fundamental romanticism of Hammerstein's tale, and the exotic canvas against which both were played stamped the show as consummate operetta.

The partners' next three shows were closer to musical comedy, despite the presence of Helen Traubel in one of them. For their last work, however, they returned to operetta. Howard Lindsay and Russell

Crouse, their librettists, set their story in Austria, the very home of operetta. The specifics of the plot, based on the true story of the Trapp family's flight from the Nazis, were anything but frivolous and maintained the modern musical play's somber undercurrents. But in its sentimentality, its romance, and its often soaring melodies ("Climb Every Mountain" and the title song), *The Sound of Music* was unalloyed, masterful operetta. (Of course, Rodgers's score also included lighter songs such as "Do Re Mi," "My Favorite Things," and the gentle "Edelweiss.")

So evident were *The Sound of Music*'s allegiances to operetta that Atkinson could not overlook them. They cued another attack. "It is dissappointing," he concluded in his review, "to see the American musical stage succumbing to the cliches of operetta." The public succumbed, too. The stage version was tremendously successful, and the film version, with Julie Andrews as the lead, became the biggest-grossing motion picture up to its time.

12

Modern American Operetta: The Expansion of the Musical Play

Rodgers and Hammerstein dominated the operetta stage as the prime movers in the first years of this epoch. But they were far from alone. A number of figures, many more often associated with revues or musical comedies, tried their hands with varying degrees of success. Arthur Schwartz composed a magnificent score for *A Tree Grows in Brooklyn* as well as the scores for *By the Beautiful Sea*, *The Gay Life*, and *Jennie*. Harold Arlen wrote an unsuccessful score for *Saratoga*, along with scores for two musicals that straddled the fence between musical comedy and operetta, *House of Flowers* and *St. Louis Woman*. Harold Rome contributed the impressive score for *Fanny*, and John Kander for *Cabaret*. Jerome Moross put music to John Latouche's clever lyrics for *The Golden Apple*. Mitch Leigh gave us *Man of La Mancha*.

The first and last of these shows—*A Tree Grows in Brooklyn* and *Man of La Mancha*—are typical of all except *The Golden Apple*. Both dealt with the shattering of escapist dreams. But the similarity ended there. *A Tree Grows in Brooklyn*, based on Betty Smith's best-selling novel, couched its sad story in sweet terms. The plot centered on the heroine's star-crossed romance with a ne'er-do-well who, despite his sincere promises, can never remain employed or sober. Intermeshed with this was a tale of the equally unsuccessful love life of a wistfully comic aunt, who in her youth had unwittingly entered into a bigamous marriage. She nonetheless longs for her faithless husband until he one day reappears and she sees him with new eyes. Schwartz mixed strong, moving ballads with his heel-kicking polka and captivating hurdy-gurdy waltz. From beginning to end the score was superbly lyrical, and its lyricism was firmly tied to popular, traditional light opera forms. Dorothy Fields's humorous, colloquial lyrics may have belonged more to musical comedy than to operetta, but again the basic story and score clearly subscribed to demands of the modern musical play.

Man of La Mancha, based on Cervantes's *Don Quixote*, took an opposing tack. It was tough, harshly cynical, and often fiercely impassioned. Of necessity it went to Spanish art forms for much of its music, but even there it was relatively free in the shaping of its melodic lines and structures. Without being ugly itself, the music successfully conveyed much of the ugliness in the story—the sexual assaults, the personal cruelty. And at times it was almost operatic in nature, demanding superior voices which, in the original production, it received. One lesser song, "The Impossible Dream," became popular immediately. In the short run at least the show has proved durable, providing its original star, Richard Kiley, with a dependable meal ticket.

Because *The Golden Apple* was so out of the mainstream, it deserves a brief word. An imaginative reworking of the *Odyssey* set in turn-of-the-century Midwestern America, it was unique in several respects. For one, it was almost entirely sung; there was little or no dialogue. Jerome Moross's clever score caught a period flavor while remaining totally contemporary. More importantly, it set aside the earnest romanticism and even the fashionable social commentary of modern musical plays, relying entirely on the unflagging wit of John Latouche's lyrics and a charming stylization to carry it over.

Yet however successful some of these works were, their authors were not the most productive advocates of modern-day operetta. The great names of this latest era, aside from Rodgers and Hammerstein, were Kurt Weill, Frederick Loewe, Leonard Bernstein, Frank Loesser, Stephen Sondheim, and Jerry Bock for music; and Alan Jay Lerner and Stephen Sondheim for books and/or lyrics.

Let's look at the composers first. Their backgrounds were not dissimilar. Except for some very early masters such as Sullivan and De Koven, great Protestant composers ignored operetta. The last important Catholic, in a glittering line from Strauss through Herbert, had been Rudolf Friml. To a man, the latest generation of composers was Jewish, and largely German-Jewish at that. Only Weill and Loewe were European-born. The others' ties to a European past varied, depending to some small extent on whether they were first- or second-generation Americans. All came from essentially middle-class homes, some prosperous, some less so. Certainly none came from the ghetto beginnings that colored the writings of an older order of Tin Pan Alley composers. Without exception they came from homes where musical arts were cherished and encouraged. Weill's father had been a cantor; Loewe's was one of the earliest Continental Danilos in *The Merry Widow;* Loesser's father taught piano. Except for Loesser, all had received substantial formal musical training. Their credentials, then, were of the highest stripe.

So was their music. Little of it was instantly recognizable as direct heir to the older schools. One striking reason was that the high romantic moment of the evening was no longer delegated to the waltz, which had virtually become operetta's hallmark since Viennese days. Time and again composers turned to the contemporary ballad when they reached crucial, impassioned love songs. However boy met girl, the two customarily courted in 4/4 time. To this extent the modern American musical play drew inspiration from its musical comedy and Tin Pan Alley competitors. Yet even here it must be recalled that "They Didn't Believe Me," the love ballad that first made all this possible, came from the pen of Jerome Kern, who had pioneered this same school of American operetta. His examples in *Show Boat,* "Make Believe," and "Why Do I Love You?," consciously or not, secured the ballad pride of place in later such operettas and thereby made possible "People Will Say We're In Love," "Some Enchanted Evening," and the host of brilliant masterpieces that followed.

Demoted but certainly not discarded, the waltz continued to thrive. It remained too expressive, too enlivening to disappear. Only its function was altered. The waltz's irresistible lilt gave voice to jubilation at life, at one's self, and even at love—*Oklahoma!*'s "Oh, What A Beautiful Mornin'!," *West Side Story*'s "I Feel Pretty," *South Pacific*'s "I'm In Love With A Wonderful Guy." And, in contrast, it was used to explore bitterness, frustration, or sadness as in *South Pacific*'s "This Nearly Was Mine" or *The King and I*'s "Hello, Young Lovers." (That so many of these waltzes are Rodgers's suggests his special affinity for the form and his link with older traditions.) If lovers no longer clasped hands in a waltz, neither did they often waltz to it. It had become something of a musical apostrophe, a medium of expression far more than an invitation to a dance. Nevertheless, its virtues and potentials were compelling enough to prompt Sondheim to write *A Little Night Music* entirely in variations of 3/4 time.

In one guise or another, many other traditional forms served these new operettas as well as they had served the older ones. No booming male chorus clanked tankards in a rousing drinking song—though one should not forget *Fiddler on the Roof*'s "To Life"—but it could raise *South Pacific*'s audiences out of their seats exulting "There Is Nothing Like A Dame." A polka, like many a waltz in other shows, could turn into a paean of joy, although in the case of *A Tree Grows in Brooklyn*'s "Look Who's Dancing" it did lead into a high stomping dance. Even the tango, which did not reach the boards until several years after *The Merry Widow*, found fresh, witty, and exciting employment in *My Fair Lady*'s "The Rain In Spain." And while Bernstein called for an even more modern beguine tempo in *West Side Story*'s "Tonight," the song's subject matter and glowing ardor made it as perfect a serenade as Romberg's in *The Student Prince*.

That these song forms were also used in musical comedy in no way obviated their legitimacy in operetta, for while the forms may have been similar, the textures were usually far richer in operetta. Limited ranges and choppy musical lines gave way to more expansive sustained reaches. Witness, for example, the eloquent flow and octave and a half spread of "Tonight." In keeping with contemporary requirements, however, both flows and spreads were more restricted than they once had been. And pyrotechnic effects achieved by trills and dangerous jumps all but totally disappeared. Moreover, old or new forms were invested with contemporary colorings. The smokier chromatics and harmonies prevalent in all other modern music prevailed here, too. Even apparent dissonances were never shunned. Familiarity quickly made them seem natural and restful, as *South Pacific*'s "Bali Ha'i" so cogently proved.

Though employed in the same arena and using the same tools and materials, most of the greater composers brought distinguishing characteristics to their work. As if to establish a clear-cut break with his earlier career, Rodgers's music from *Oklahoma!* on, while constantly innovative and progressive, turned to a four-square directness, a quiet sincerity, and a carefully controlled sentimentality. The same could be said of Bock, perhaps the most underrated of the recent masters. His best music often had that elusive charm, that "odor of sachet," that Kern had set as a goal. Loewe was undoubtedly the most traditional and derivative of the group. Fortunately, he coupled a bottomless well of melody with an uncanny ability to catch sounds of place and time. Weill and Bernstein were especially interesting for their obvious debts to jazz. While both were expressively lyrical, they allowed jazz's steely nuances and sardonic undercurrents ample play. Bernstein also brought an infectious sense of humor to much of his composition. But the wittiest and most offbeat music was Loesser's. For all its breadth and timbre, his songs often had a colloquial immediacy that was unique among his peers.

Unquestionably the most intriguing of the major composers was Sondheim, as much for the unanswered questions his works raised as for his obvious, if qualified, triumphs. He brought an exceptionally clever mind and lavish technical gifts to his writings. Posterity may well hail—or condemn—him as father of the conceptualized musical wherein stories and characters seem to arise from attitudes and perceptions rather than the other way around. His music and lyrics (Sondheim is the logical figure to carry us from composers to wordsmiths) and the very stories he presented to his librettists probed with depth and sharpness the minds of the figures who passed across his stages and the worlds they inhabited. He seemed able to grasp and convey musically fleeting pictures that passed through his characters' subconscious. What he was incapable of or, given his huge talent, more likely unwilling to do was to project these thoughts and feelings in musical terms that audiences could retain and repeat. Even an exquisitely conceived song such as "Send In The Clowns" presented difficulties for the man in the street who might remember it and whistle it casually. Sondheim's songs are the hardest to extract from their context and, therefore, the hardest to grow affectionate about.

Sondheim brought to his lyrics the same virtues and shortcomings. Some of his lyrics are the most brilliant our popular musical stage has ever heard, packed as they are with matter and ingenious rhyme. At times, unfortunately, they seem to wear their ingenuity on their sleeve. Moreover, Sondheim's dark views of life prompted him frequently to examine its uglier aspects in naked terms. Commonplaces such as "I love you," commonplaces that nevertheless help give songs widespread currency, were rare in his later efforts. Sondheim's librettist, Hugh Wheeler, skillfully caught both Sondheim's intent and approach.

Alan Jay Lerner, every bit as literate and witty as Sondheim, worked on a far more elegant level. High style and grace were his keynotes, whether he was dealing with medieval kings, Edwardian patricians, or rough-and-tumble prospectors. His public view of life was far more traditional than Sondheim's. As his own librettist, Lerner's sense of construction sometimes created obstacles, but his dialogue was always superior and never showy.

All that can be said of Oscar Hammerstein has assuredly been said before. His admirers point to his unostentatious professional craftsmanship, his ability to re-create the spontaneity and honesty of everyday speech, and the often incandescent poetry of his better lyrics. His detractors point to a lack of easy laughter in much of his work, an occasional overreaching simile in his lyrics, and a penchant for treacly sentimentality. On balance, his assets no doubt more than compensate for his failings. And it must be remembered that he was not the sole librettist in his next-to-last operetta and that in the final one he left the libretto to others.

Having looked at the writers, let us examine a few of their major works that have not yet been fully discussed. Lerner and Loewe's masterpiece, *My Fair Lady*, demands pride of place. In one respect, like *The Golden Apple*, it was out of the musical mainstream in that its salient virtues were its incomparably literate wit and its unwavering stylish elegance. It stood in perceptible contrast not only to its own generation of operettas but also to the only other operetta that had ever been based on a George Bernard Shaw play, *The Chocolate Soldier*. *The Chocolate Soldier* was a Viennese piece, created at the height of the Viennese waltz-operetta craze. It retained the basic story of a pacifist fighting-man, but replaced Shaw's high wit with buffoonery. *My Fair Lady*, on the other hand, based on *Pygmalion*, matched Shaw's style, tone, and intellectualism from its first curtain to its last. Professor Higgins once again sets out to teach a guttersnipe how to pass for a lady, and once again receives some unsought for lessons in return.

Loewe's score was not strikingly original, but it had, in an impressionistic way, an evocative sense of period. The show's principal love song, "On The Street Where You Live," was a ballad, not, as it would have been in the show's period, a waltz. Even the heroine's exultant response to an evening of waltzing, "I Could Have Danced All Night," was in 4/4 time. Of the show's major numbers only "Show Me" was in 3/4 time, and that was hardly a standard waltz. But Loewe put the tango, new in the story's period, to remarkable use in "The Rain In Spain" and brought in British music-hall mannerisms for Doolittle's "With A Little Bit Of Luck" and "Get Me To The Church On Time." For much of Higgins's material, Lerner and Loewe returned the patter song to the operetta stage.

But when the team turned to the Arthurian legend for their next show, *Camelot*, they met with disappointment, though by no means with failure. Time has suggested that *Camelot*'s greatest failing was that it did not live up to unrealistic expectations. Certainly Loewe's lush, varied score was not at fault, nor were Lerner's deft lyrics. Indeed, the show's ballad, "If Ever I Would Leave You," enjoyed an instant, wide-ranging vogue. Some of the disappointment must have derived from the libretto, especially the uncertain, semi-comic handling of the villainous Modred. In the original production Modred came across as a spoiled brat, hardly of the right tenor for the high romance in the rest of the operetta. While the show was on its post-Broadway tour, President Kennedy was assassinated. The brief cultural flowering his administration had bravely fostered seemed nipped in the bud. Lerner's line about Camelot's "one brief shining moment" took on a poignant significance that nonetheless called attention to the show again and allowed many to reconsider its merit.

Earlier on, in 1951, the team had offered a more traditional musical play, set in the ever popular Gold Rush days. *Paint Your Wagon* centered upon the efforts of an aging, hard-drinking settler to keep his world from falling apart after gold is discovered on his land, and to protect his daughter from the cynicism of those around her. After the gold peters out and the old man dies, his daughter and her miner beau determine to establish a farm on the land. Loewe's colorful score included two soft, pathetic numbers, "I Still See Elisa" and "Wandrin' Star," the stirring "They Call The Wind Maria," the thumping title song, and a beguinelike ballad, "I Talk To The Trees."

Earlier still, in 1947, Lerner and Loewe had successfully explored another possibility for operetta, the realm of pure fantasy. For his *Brigadoon* libretto, Lerner had resorted to an ancient folk motif, a village that disappears only to return at regular intervals. The little Scottish village of Brigadoon comes back to life for a day once every hundred years. Two lost American hunters stumble upon it during its latest resurrection. They fall in love with village girls, and the love of one of the men proves so strong that when he leaves the hamlet and comes back months later to seek it out, it miraculously appears just long enough for him to reenter it. For so exotically romantic a tale, Loewe composed a tender score. Songs such as "Come To Me, Bend To Me," "Almost Like Being In Love," "The Heather On The Hill," "Waitin' For My Dearie," and "I'll Go Home With Bonnie Jean" were not only enchantingly atmospheric, but were lyrical and broad enough to classify as operetta writing. Although *Brigadoon* suggested an alternative to bygone Americana as inspiration for operetta, fantasy was too tricky a device to be employed with any regularity.

Lerner and Loewe were also responsible for the most important film operetta of the era. *Gigi* was a musical version of a stage play drawn in turn from a story by Colette. That story described the coming of age of a young Parisian girl at the turn of the century. Leslie Caron, Louis Jourdan, Maurice Chevalier, and Hermione Gingold sang and danced a succession of fetching songs that ranged from the appropriately bubbly "The Night They Invented Champagne" to the wry, wistful "I Remember It Well" and from the jauntily boulevardier "Thank Heaven For Little Girls" to the haunting, musically long-lined title song.

As far back as 1938 with *Knickerbocker Holiday*, and possibly with parts of *Johnny Johnson* in 1936, Kurt Weill had hinted at forthcoming directions in modern American operetta. Despite some superior music, Weill's 1945 operetta, *The Firebrand of Florence*, was a quick failure. So was his 1947 opera, *Street Scene*, although it has since found a place in opera repertories. His last work, *Lost in the Stars*, has found a similar niche, albeit the piece straddles the fence between pure opera and Broadway far more than *Street Scene*. Weill and Maxwell Anderson (who had collaborated on *Knickerbocker Holiday*) took their text from Alan Paton's poetically eloquent tale of South African racial injustice, *Cry, the Beloved Country*. Weill consciously rejected using African musical idioms, and instead borrowed from American Negro and American Western lyricism as well as recalling some of his own earlier, German-filtered jazz. Musically the operetta was moving and memorable. But Anderson kept shifting his story from the direct action at center stage to a sort of Greek chorus on the sidelines. The result was a choppy, often static, and ultimately unsatisfying book.

An operetta with a similarly eclectic score and not dissimilar book problems was Leonard Bernstein's *Candide*. Lillian Hellman adapted Voltaire's novella, retaining its episodic nature but missing its rapier wit and elegant, airy hauteur. John Latouche, Richard Wilbur, and Dorothy Parker provided lyrics that were sometimes brilliant and original while just as often Broadway-like and banal. Bernstein was the show's hero, leading Brooks Atkinson to proclaim "the joyous variety, humor, and richness of the score." *Candide*'s glittering overture soon became a concert favorite. It remains a brilliant composition, closer in feeling to the carefully structured and textured overtures from long-gone comic operas and opéra bouffes than to the hastily assembled medleys that now generally pass for overtures. In the show itself, neglected forms such as the mazurka, the schottische, and the gavotte were vibrantly resurrected along with the waltz and the tango. Much of the material was operatic in nature. Luckily the original cast included Barbara Cook, Robert Rounseville, and Irra Petina, who could handle Bernstein's most challenging passages seemingly with ease. (A 1974 revival had a new book by Hugh Wheeler and additional lyrics by Stephen Sondheim and was staged in electrifying fashion by Harold Prince. It was the musical hit of the season and brought further acclaim to Bernstein's brilliant score.)

In 1957, with *West Side Story*, Bernstein moved from the cruel, luxurious world of eighteenth-century aristocracy to the cruel, shabby world of modern ethnic ghettos. Such a setting was, moreover, a marked turnabout from the palaces and Italian Renaissance byways in which the action originally unfolded, for *West Side Story* was a twentieth-century American redaction of Shakespeare's *Romeo and Juliet*, with the feuding Capulets and Montagues giving place to uneducated, unmanageable Puerto Ricans who come into conflict with low-life natives. There is little point here in dwelling on just how literate and moral the first families of the Italian Renaissance were. What mattered for theatrical purposes was the general perception of a beautifully ordered, richly comparisoned world in which noble stories could be played out. In *West Side Story*, librettist Arthur Laurents, lyricist Stephen Sondheim, choreographer Jerome Robbins (whose conception was the genesis of the show), and Bernstein all courageously eschewed such a sheltering panoply. They retold the centuries-old tale with what for the time was stark, sometimes offensive, realism. They set their retelling in dilapidated, dangerous ghetto alleys and shabby tenements, and dwelled, almost lovingly, on the meanness, cruelty, tastelessness, and hypocrisy of their figures. (A revival of the show nearly a quarter-century after its premiere, however, found several critics surprised at its relative tameness.) If the basic tale and Robbins's electrifying dances gave *West Side Story* its immediate theatricality, the earnestness of the retelling, Sondheim's often poetic lyrics, and Bernstein's exceptionally lyric score gave it its real inner strength and classified it unquestionably as operetta. The haunting beauty and breadth of songs such as "One Hand, One Heart," "Maria," and "Tonight" helped soothe and reassure those made uncomfortable by the musical's disturbing aspects.

Frank Loesser tried to encompass the whole contemporary American musical theatre in *The Most Happy Fella*—and almost succeeded. The show worked effectively on stage, although it ricocheted from musical comedy to operetta to almost grand opera. Quite possibly Loesser had doubts about either his ability to create genuinely serious musicals or his public's willingness to receive them. Certainly his best musical comedies—*Guys and Dolls* and *How To Succeed in Business Without Really Trying*—had a stylistic or tonal integrity. In the case of *The Most Happy Fella*, his mixing of genres represented an unmistakable decision on his part, since musical comedy numbers such as "Standing On The Corner" and "Big D" were given to frivolous characters who didn't exist in Sidney Howard's *They Knew What They Wanted*, the play on which the show was based. The principal story concerned the quirky courtship of a young waitress by an aging immigrant Italian wine-maker. In the girl's confusion when she realizes how old her suitor is, she has an affair with a farmhand of her own age. The Italian later forgives her, and in the end they wed. Quite possibly Loesser planned from the start to cast an opera basso as the Italian (Robert Weede played it originally, and Giorgio Tozzi in the recent revival), for the most operatic music was given to the character. It was music overflowing with Italian fervor and warmth. The girl and her farmhand lover had more theatrically restricted material, music in the fashion of the contemporary musical play. Loesser's second operetta, *Greenwillow*, had a more consistent tone, but may have, in a sad way, confirmed whatever reservations or fears Loesser held about working with the genre, for it proved a rather watery theatrepiece. A bucolic portrait of a young man's coming of age, the show nonetheless retains a coterie of admirers.

Jerry Bock and his lyricist, Sheldon Harnick, gave Broadway two operetta masterworks, *She Loves Me*, to a libretto by Joe Masteroff, and *Fiddler on the Roof*, to a libretto by Joseph Stein. The settings harked back to those beloved in waltz-operetta and comic opera's heyday—Middle Europe and Russia. But it was not to the grand palaces and upper crust cafés that the authors turned. *She Loves Me*, based on a Hungarian play of the 1930s, unfolded in the world of petit bourgeoisie and dealt with the romance of two clerks in a small boutique. *Fiddler on the Roof*, based on Sholom Aleichem's stories about Yiddish life in Russia, took place entirely in a shabby Jewish shtetl. It depicted the difficulties a pious dairyman encounters in trying to raise a family, difficulties that ultimately force him and his family to flee their beloved home and emigrate to America. *She Loves Me* was pure escapist romanticism, of a sort modern-day operetta rarely employed. *Fiddler on the Roof* was certainly less escapist and less romantic. Although it was thoroughly sentimental, the cruelties so inextricably interwoven into its plot made the sentimentality believable and palpable. The original productions of both musicals benefited from exceptional performances by their stars: Barbara Cook in *She Loves Me* and Zero Mostel in *Fiddler*.

Bock and Harnick supplied songs that in each case unerringly caught both period and tone. The score for *She Loves Me* was crammed with charming bonbons, including a delicious waltz, "Dear Friend," the comic, exulting "Ice Cream," and the exuberant title song. *Fiddler*'s score, calling so often on Jewish strains of melancholy and compassion, was richly melodic. Songs as diverse as "Matchmaker, Matchmaker," "To Life" ("L'Chaim"), and "Sunrise, Sunset" quickly became standards. *Fiddler* surpassed both *Oklahoma!*'s and *My Fair Lady*'s stands to become Broadway's longest-run musical when it closed.

Disillusionment was the keynote of Stephen Sondheim's two great operettas, *A Little Night Music* and *Sweeney Todd*. In the first, this disillusionment manifested itself in a romantic world-weariness; in the latter it was transformed into bitterness and unfettered hatred. *A Little Night Music* was set in turn-of-the-century Sweden and played out against a background of steely northern moonlight and flickering candles. The plot revolved about the unsatisfactory love lives of a middle-aged lawyer, his child-bride, his ex-mistress and her new liaison, and an ancient grand dame. The entire score was composed in variations of 3/4 time. Except for the half-sad, half-mocking "Send In The Clowns," the score had little popular appeal but worked effectively in the theatre to evoke the moods and sentiments of the characters. *Sweeney Todd*'s music was harsher and more dramatic, reflecting its gruesome tale of a vengeful, murderous barber who gives his corpses to his lady love to bake into pies. Nothing from Sondheim's score proved popular away from the playhouse, but his acute ear for musical characterization and his incomparable lyrics have helped make the show stunning theatre.

The leading performers in these operettas, vocally at least, represented a wildly mixed bag. Films, radio, and nightclubs had long since proven profitable lures, and early on in this last period, television had appeared with its own powerful inducements to desert the stage. Stars of truly operatic quality had all but vanished, and many of those who did come to Broadway, great artists such as Ezio Pinza, Robert Weede, Robert Rounseville, Irra Petina, and Patrice Munsel, did so only after their operatic careers had peaked or faded. Some fine singers with exceptional stage voices were able to make notable careers by singing both operetta and musical comedy. Alfred Drake was unquestionably the greatest of these—not only a magnificent baritone, but a riveting actor and brilliant high comedian. John Raitt, Richard Kiley, and Barbara Cook fell into this category, and there are many who believe Miss Cook, had she chosen, could have had a career in opera. Distinguished actors and actresses also occasionally cast their lot with operetta and enjoyed notable triumphs, particularly Rex Harrison and Shirley Booth. From the 1950s on, miking and amplification helped any number of weaker-voiced performers project over the heavy orchestrations that had come to prevail.

The onset of these more somber musical plays and the attendant ballyhoo they were awarded had both beneficial and harmful effects on the musical theatre. In their artfulness and seriousness they gave a new richness to our musical stage and even allowed opera, or something close to it, entry to the Broadway arena. Works such as Weill's *Street Scene*, Marc Blitzstein's *Regina*, and Gian Carlo Menotti's efforts received a welcome—or at least a hearing—they might not otherwise have gained. However, the almost ubiquitous adulation accorded these modern operettas often went hand in hand with a vitriolic belittling of the older schools. To some extent this was inevitable. Each generation praises its own achievement at the expense of the past. It is an unfortunate phenomenon, not merely in the theatre. In time, this most recent school may well undergo a temporary decline as still newer treatments and newer ideas take hold and have their day. Historically, however, the rejected past is reevaluated at intervals. When, in the long run, a complete reevaluation of our musical stage occurs it may well bring back works by Sousa, Herbert, Romberg, Friml, and Kern in active repertory with those of Rodgers and Hammerstein, Lerner and Loewe, Bernstein, and Sondheim.

Appendix

Principals and Credits for Important Operettas

This appendix provides brief statistics on a selected list of operettas discussed in the book. For the most part, selections are the obvious ones, but in a few instances they are admittedly arbitrary. For some of the earlier importations, rarely or never revived in English, the original French or German song titles have been listed.

La Grande Duchesse de Gérolstein. Music by Jacques Offenbach; libretto by Henri Meilhac and Ludovic Halévy. French Theatre. September 24, 1867. Lucille Tostée (The Grand Duchess), M. Guffroy (Fritz), M. Duchesne (General Boum). Principal songs: "Voici le sabre de mon père," "Ah! que j'aime les militaires," "Dites-lui qu'on l'a remarqué." 165 performances.

H.M.S. Pinafore. Music by Arthur Sullivan; libretto by W. S. Gilbert. Standard Theatre. January 15, 1879. Thomas Whiffen (Sir Joseph Porter), Eugene Clark (Captain Corcoran), Henri Laurent (Ralph Rackstraw), Eva Mills (Josephine), Blanche Galton (Buttercup). Principal songs: "We Sail The Ocean Blue," "I'm Called Little Buttercup," "I Am The Captain Of The 'Pinafore,' " "When I Was A Lad." 175 performances.

Fatinitza. Music by Franz von Suppé; libretto by F. Zell and Richard Genée. Germania Theatre. April 14, 1879. Frl. Kuhse (Wladimir), Herr Franosch (Graf Kantshakoff), Frl. Kuster (Lydia). Principal songs: "Wie schade wie schade," "Welche Lust beim Spiel," "Abends wenn vom Minaret." 14 performances in German; 148 in English.

The Pirates of Penzance. Music by Arthur Sullivan; libretto by W. S. Gilbert. Fifth Avenue Theatre. December 31, 1879. Hugh Talbot (Frederick), J. H. Ryley (Major-General Stanley), Blanche Roosevelt (Mabel). Principal songs: "Poor Wandering One," "I Am The Very Model Of A Modern Major-General," "With Catlike Tread." 72 performances.

The Vicar of Bray. Music by Edward Solomon; libretto by Sydney Grundy. Fifth Avenue Theatre. October 2, 1882. Harry Allen (The Vicar), George Olmi (Tommy Morton), L. Cadwallader (Rev. Henry Sanford), Marie Jansen (Dorothy), Edith Bland (Nelly Bly), Jennie Hughes (Mrs. Morton). Principal songs: "I'm As Sharp As A Ferret," "Tell Me True, Love," "Come Back To Me," "O William, Sweet William." 8 performances.

The Mikado. Music by Arthur Sullivan; libretto by W. S. Gilbert. Fifth Avenue Theatre. August 19, 1885. [Actually, two unauthorized productions had appeared in New York earlier—one on July 20, the other on August 10.] F. Federici (The Mikado), Courtice Pounds (Nanki-Poo), F. Billington (Pooh-Bah), George Thorne (Ko-Ko), Geraldine Ulmar (Yum-Yum), Elsie Cameron (Katisha). Principal songs: "A Wandering Minstrel, I," "The Flowers That Bloom In The Spring," "Willow, Tit-willow." 250 performances.

The Little Tycoon. Music and libretto by Willard Spenser. Standard Theatre. March 29, 1886. R. E. Graham (General Knickerbocker), Carrie M. Dietrich (Violet), W. S. Rising (Alvin), E. H. Van Veghten (Lord Dolphin). Principal songs: "The Cats On Our Back Fence," "Love Comes Like A Summer Sigh." 108 performances.

Erminie. Music by Edward Jakobowski; libretto by Harry Paulton. Casino Theatre. May 10, 1886. Pauline Hall (Erminie), Francis Wilson (Cadeaux), W. S. Daboll (Ravennes). Principal songs: "Lullaby" ("Dear Mother, In Dreams I See Her"), "We're A Philanthropic Couple." 571 performances.

The Begum. Music by Reginald De Koven; libretto by Harry B. Smith. Fifth Avenue Theatre. November 21, 1887. De Wolf Hopper (Howja-Dhu), Mathilde Cottrelly (The Begum), Hubert Wilke (Klahm-Chowdee), Jefferson De Angelis (Jhustt-Naut). Principal songs: "Hear Ye The Birds," "Do You Remember?," "When War Began." 24 performances.

Robin Hood. Music by Reginald De Koven; libretto by Harry B. Smith. Standard Theatre. September 28, 1891. Tom Karl (Robin Hood), Eugene Cowles (Will Scarlet), Caroline Hamilton (Maid Marian), Jessie Bartlett Davis (Alan-a-Dale), W. H. MacDonald (Little John), Henry Clay Barnabee (Sheriff of Nottingham). Principal songs: "Brown October Ale," "Oh, Promise Me." 72 performances (in two slightly separated engagements).

El Capitan. Music by John Philip Sousa; libretto by Charles Klein; lyrics by Sousa and Tom Frost. Broadway Theatre. April 20, 1896. De Wolf Hopper (Medigua), Charles Klein (Pozzo), Edna Wallace Hopper (Estrelda). Principal songs: "El Capitan's Song," "Sweetheart, I'm Waiting," "A Typical Tune Of Zanzibar." 112 performances.

The Fortune Teller. Music by Victor Herbert; book and lyrics by Harry B. Smith. Wallack's Theatre. September 26, 1898. Alice Nielsen (Musette and Irma), Eugene Cowles (Sandor), Frank Rushworth (Ladislas), Joseph Herbert (Count Berezowski), Marguerite Sylva (Pompon), Joseph Cawthorn (Boris). Principal songs: "Gypsy Love Song," "Romany Life," "Only In The Play," "Always Do As People Say You Should." 40 performances.

When Johnny Comes Marching Home. Music by Julian Edwards; book and lyrics by Stanislaus Stange. New York Theatre. December 16, 1902. Zetti Kennedy (Kate), William G. Stewart (John). Principal songs: "My Own United States," "My Honeysuckle Girl," "Fairyland." 71 performances.

The Prince of Pilsen. Music by Gustav Luders; book and lyrics by Frank Pixley. Broadway Theatre. March 17, 1903. John W. Ransome (Hans Wagner), Albert Parr (Tom), Lillian Coleman (Nellie), Arthur Donaldson (Prince Carl Otto). Principal songs: "The Heidelberg Stein Song," "The Message Of The Violet." 143 performances.

Mlle. Modiste. Music by Victor Herbert; book and lyrics by Henry Blossom. Knickerbocker Theatre. December 25, 1905. Fritzi Scheff (Fifi), Walter Percival (Capt. Etienne de Bouvray), William Pruette (Henri de Bouvray), Claude Gillingwater (Hiram Bent). Principal songs: "Kiss Me Again" [actually part of a longer number, "If I Were On The Stage"], "I Want What I Want When I Want It," "The Mascot Of The Troop," "The Time And The Place And The Girl." 202 performances.

The Merry Widow. Music by Franz Lehar; libretto by Victor Leon and Leo Stein. English adaptation by Basil Hood and Adrian Ross. New Amsterdam Theatre. October 21, 1907. Ethel Jackson (Sonya), Donald Brian (Danilo), R. E. Graham (Popoff), William Weedon (Camille). Principal songs: "The Merry Widow Waltz" ("I Love You So"), "Maxim's," "Vilja," "Women." 416 performances.

A Waltz Dream. Music by Oscar Straus; libretto by Felix Doermann and Leopold Jacobson. English adaptation by Joseph W. Herbert. Broadway Theatre. January 27, 1908. Edward Johnson (Nicki), Magda Dahl (Princess Helen). Principal songs: "Love's Roundelay," "Piccolo," "Life Is Love And Laughter." 111 performances.

The Dollar Princess. Music by Leo Fall; libretto by Arthur Maria Willner and Fritz Grünbaum. English adaptation by George Grossmith, Jr., and Adrian Ross. Knickerbocker Theatre. September 6, 1909. Valli Valli (Alice), Donald Brian (Freddy). Principal songs: "The Dollar Princesses," "My Dream Of Love." 288 performances.

The Chocolate Soldier. Music by Oscar Straus; libretto by Rudolph Bernauer and Leopold Jacobson. English adaptation by Stanislaus Stange. Lyric Theatre. September 13, 1909. J. E. Gardner (Bumerli), Ida Brooks Hunt (Nadina), William Pruette (Popoff). Principal songs: "My Hero," "Sympathy," "Forgive, Forgive," "The Bulgarians," "Thank The Lord The War Is Over." 296 performances.

Naughty Marietta. Music by Victor Herbert; book and lyrics by Rida Johnson Young. New York Theatre. November 7, 1910. Emma Trentini (Marietta), Orville Harrold (Dick), Marie Duchene (Adah), Edward Martindel (Etienne). Principal songs: "Tramp, Tramp, Tramp," " 'Neath The Southern Moon," "Italian Street Song," "I'm Falling In Love With Someone," "Ah, Sweet Mystery Of Life." 136 performances.

The Pink Lady. Music by Ivan Caryll; book and lyrics by C. M. S. McLellan. New Amsterdam Theatre. March 31, 1911. Hazel Dawn (Claudine), William Elliot (Lucien Garidel), Alice Dovey (Angele), Frank Lalor (Dondidier). Principal songs: "My Beautiful Lady" (also sung as "The Kiss Waltz"), "Love Is Divine," "By The Saskatchewan," "Donny Didn't, Donny Did." 312 performances.

The Firefly. Music by Rudolf Friml; book and lyrics by Otto Harbach. Lyric Theatre. December 2, 1912. Emma Trentini (Nina), Craig Campbell (Jack), Audrey Maple (Geraldine), Melville Stewart (John). Principal songs: "Giannina Mia," "When A Maid Comes Knocking At Your Heart," "Love Is Like A Firefly," "Sympathy." 120 performances.

Sweethearts. Music by Victor Herbert; book by Harry B. Smith and Fred De Gressac [Mrs. Victor Maurel]; lyrics by Robert B. Smith. New Amsterdam Theatre. September 8, 1913. Christie MacDonald (Sylvia), Thomas Conkey (Franz), Edwin Wilson (Lt. Karl), Tom McNaughton (Mikel Mikeloviz). Principal songs: "Sweethearts," "Every Lover Must Meet His Fate," "Jeanette And Her Little Wooden Shoes," "Pretty As A Picture." 136 performances.

Sari. Music by Emmerich Kalman; libretto by Julius Wilhelm and Fritz Grünbaum. English adaptation by C. C. S. Cushing and E. P. Heath. Liberty Theatre. January 13, 1914. Mizzi Hajos [later simply Mitzi] (Sari), Charles Meakins (Gaston), Van Rensselaer Wheeler (Racz), Harry Davenport (Cadeaux). Principal songs: "Time, Oh Time You Tyrant," "Love Has Wings," "My Faithful Stradivari," "Love's Own Sweet Song." 151 performances.

Maytime. Music by Sigmund Romberg; book and lyrics by Rida Johnson Young. Shubert Theatre. August 16, 1917. Peggy Wood (Ottilie), Charles Purcell (Richard). Principal songs: "Will You Remember?," "The Road To Paradise," "Jump Jim Crow," "In Our Little Home Sweet Home." 492 performances.

Apple Blossoms. Music by Victor Jacobi and Fritz Kreisler; book and lyrics by William LeBaron. Globe Theatre. October 7, 1919. Wilda Bennett (Nancy), John Charles Thomas (Philip), Percival Knight (Richard), Florence Shirley (Anne), Roy Atwell (Harvey), Fred Astaire (Johnny), Adele Astaire (Molly). Principal songs: "You Are Free," "Little Girls, Goodbye," "Who Can Tell?" [this melody was later reemployed in a film for the song "Stars In Your Eyes"]. 256 performances.

Blossom Time. Music by Franz Schubert, adapted by Sigmund Romberg; book and lyrics by Dorothy Donnelly (based on a Viennese operetta). Ambassador Theatre. September 29, 1921. Bertram Peacock (Schubert), Olga Cook (Mitzi), Howard Marsh (Schober). Principal songs: "Song Of Love," "Tell Me Daisy," "Serenade." 592 performances.

Rose-Marie. Music by Rudolf Friml and Herbert Stothart; book and lyrics by Otto Harbach and Oscar Hammerstein II. Imperial Theatre. September 2, 1924. Mary Ellis (Rose-Marie), Dennis King (Jim), Arthur Deagon (Sgt. Malone), William Kent (Hard-Boiled Herman), Peal Regay (Wanda). Principal songs: "Rose-Marie," "Indian Love Call," "Song Of The Mounties," "Totem Tom-Tom." 557 performances.

The Student Prince. Music by Sigmund Romberg; book and lyrics by Dorothy Donnelly. Jolson Theatre. December 2, 1924. Howard Marsh (Karl Franz), Ilse Marvenga (Kathie), Greek Evans (Dr. Engel). Principal songs: "Deep In My Heart, Dear," "Serenade," "Golden Days," "Drinking Song." 608 performances.

The Vagabond King. Music by Rudolf Friml; book by Brian Hooker, Russell Janney, and W. H. Post; lyrics by Hooker. Casino Theatre. September 21, 1925. Dennis King (Villon), Carolyn Thomson (Katherine), Jane Carroll (Huguette). Principal songs: "Song Of The Vagabonds," "Only A Rose," "Someday," "Love Me, Tonight," "Huguette's Waltz," "Love For Sale." 511 performances.

Song of the Flame. Music by George Gershwin and Herbert Stothart; book and lyrics by Otto Harbach and Oscar Hammerstein II. 44th St. Theatre. December 30, 1925. Tessa Kosta (Aniuta), Guy Robertson (Volodyn). Principal song: "Song Of The Flame." 219 performances.

Countess Maritza. Music by Emmerich Kalman; book and lyrics by Julius Brammer and Alfred Grunwald. English adaptation by Harry B. Smith. Shubert Theatre. September 18, 1926. Yvonne D'Arle (Maritza), Walter Woolf (Tassilo), Carl Randall (Szupan), Vivian Hart (Lisa), Odette Myrtil (Manja). Principal songs: "Play Gypsies—Dance, Gypsies," "The Music Thrills Me," "In The Days Gone By," "The One I'm Looking For." 318 performances.

The Desert Song. Music by Sigmund Romberg; book by Otto Harbach, Oscar Hammerstein II, and Frank Mandel; lyrics by Harbach and Hammerstein.

Casino Theatre. November 30, 1926. Vivienne Segal (Margot), Robert Halliday (Pierre), Pearl Regay (Azuri), William O'Neal (Sid El Kar), Eddie Buzzell (Bennie). Principal songs: "The Riff Song," "The Desert Song," "Romance," "One Alone," "One Flower Grows Alone In Your Garden." 471 performances.

Rio Rita. Music by Harry Tierney; book by Guy Bolton and Fred Thompson; lyrics by Joseph McCarthy. Ziegfeld Theatre. February 2, 1927. Ethelind Terry (Rita), J. Harold Murray (Capt. Stewart). Principal songs: "Rio Rita," "The Rangers' Song," "If You Are In Love, You'll Waltz," "The Kinkajou," "Following The Sun Around." 494 performances.

My Maryland. Music by Sigmund Romberg; book and lyrics by Dorothy Donnelly. Jolson Theatre. September 12, 1927. Evelyn Herbert (Barbara Frietchie), Nathaniel Wagner (Capt. Trumbull), Warren Hull (Jack). Principal songs: "Your Land And My Land," "(The Same) Silver Moon," "Won't You Marry Me?," "Mother." 312 performances.

Show Boat. Music by Jerome Kern; book and lyrics by Oscar Hammerstein II. Ziegfeld Theatre. December 27, 1927. Norma Terris (Magnolia), Howard Marsh (Ravenal), Jules Bledsoe (Joe), Helen Morgan (Julie), Charles Winninger (Cap'n Andy). Principal songs: "Make Believe," "Ol' Man River," "Can't Help Lovin' Dat Man," "You Are Love," "Why Do I Love You?," "Bill." 572 performances.

The Three Musketeers. Music by Rudolf Friml; book by William Anthony McGuire; lyrics by Clifford Grey and P. G. Wodehouse. Lyric Theatre. March 13, 1928. Dennis King (d'Artagnan), Vivienne Segal (Constance), Joseph Macaulay (Aramis). Principal songs: "March Of The Musketeers," "Ma Belle," "Your Eyes," "My Sword And I." 318 performances.

The New Moon. Music by Sigmund Romberg; book by Oscar Hammerstein II, Frank Mandel, and Laurence Schwab; lyrics by Oscar Hammerstein II. Imperial Theatre. September 19, 1929. Evelyn Herbert (Marianne), Robert Halliday (Robert Misson), William O'Neal (Philippe). Principal songs: "Stouthearted Men," "One Kiss," "Wanting You," "Softly As In A Morning Sunrise," "Lover, Come Back To Me." 506 performances.

Bitter Sweet. Music, book, and lyrics by Noel Coward. Ziegfeld Theatre. November 5, 1929. Evelyn Laye (Sarah), Gerald Nodin (Carl), Mireille (Manon), Desmond Jeans (Capt. Lutte). Principal songs: "I'll See You Again," "Dear Little Café," "Tokay," "Zigeuner." 159 performances.

Strike Up the Band. Music by George Gershwin; book by Morrie Ryskind and George S. Kaufman; lyrics by Ira Gershwin. Times Square Theatre. January 14, 1930. Bobby Clark (Col. Holmes), Paul McCullough (Gideon), Blanche Ring (Mrs. Draper), Dudley Clements (Horace J. Fletcher), Doris Carson (Anne), Gordon Smith (Timothy). Principal songs: "Strike Up The Band," "Soon," "I've Got A Crush On You," "A Typical Self-Made American." 191 performances.

The Cat and the Fiddle. Music by Jerome Kern; book and lyrics by Otto Harbach. Globe Theatre. October 15, 1931. Georges Metaxa (Victor), Bettina Hall (Shirley), George Meader (Pompineau), Odette Myrtil (Odette). Principal songs: "The Night Was Made For Love," "Try To Forget," "She Didn't Say 'Yes.' " 395 performances.

Of Thee I Sing. Music by George Gershwin; book by George S. Kaufman and Morrie Ryskind; lyrics by Ira Gershwin. Music Box Theatre. December 26, 1931. William Gaxton (Wintergreen), Victor Moore (Throttlebottom), Lois Moran (Mary Turner), Florence Ames (French Ambassador). Principal songs: "Of Thee I Sing," "Love Is Sweeping The Country," "Who Cares?," "The Illegitimate Daughter." 441 performances.

Music in the Air. Music by Jerome Kern; book and lyrics by Oscar Hammerstein II. Alvin Theatre. November 8, 1932. Al Shean (Dr. Lessing), Walter Slezak (Karl), Natalie Hall (Frieda), Katherine Carrington (Sieglinde), Tullio Carminati (Bruno), Reinald Werrenrath (Cornelius). Principal songs: "I've Told Every Little Star," "There's A Hill Beyond A Hill," "One More Dance," "The Song Is You." 342 performances.

Oklahoma! Music by Richard Rodgers; book and lyrics by Oscar Hammerstein II. St. James Theatre. March 31, 1943. Alfred Drake (Curly), Joan Roberts (Laurey), Howard Da Silva (Jud), Celeste Holm (Ado Annie), Lee Dixon (Will). Principal songs: "Oh, What A Beautiful Mornin'," "The Surrey With The Fringe On Top," "People Will Say We're In Love," "Out Of My Dreams," "Oklahoma!" 2,212 performances.

Song of Norway. Music and lyrics by Robert Wright and George Forrest (music adapted from Grieg melodies); book by Milton Lazarus. Imperial Theatre. August 21, 1944. Lawrence Brooks (Grieg), Helena Bliss (Nina), Robert Shafer (Rikard), Irra Petina (Louisa), Ivy Scott (Grieg's mother). Principal songs: "Strange Music," "I Love You," "Freddy And His Fiddle," "Now." 860 performances.

Up in Central Park. Music by Sigmund Romberg; book and lyrics by Herbert and Dorothy Fields. Century Theatre. January 27, 1945. Wilbur Evans (John), Maureen Cannon (Rosie). Principal songs: "Close As Pages In A Book," "April Snow," "The Fireman's Bride," "The Big Back Yard." 504 performances.

Carousel. Music by Richard Rodgers; book and lyrics by Oscar Hammerstein II. Majestic Theatre. April 19, 1945. John Raitt (Billy), Jan Clayton (Julie), Jean Darling (Carrie), Eric Mattson (Mr. Snow), Christine Johnson (Nettie), Murvyn Vye (Jigger). Principal songs: "If I Loved You," "June Is Bustin' Out All Over," "What's The Use Of Wond'rin'?," "You'll Never Walk Alone." 890 performances.

Brigadoon. Music by Fredrick Loewe; book and lyrics by Alan Jay Lerner. Ziegfeld Theatre. March 13, 1947. David Brooks (Tommy), Marion Bell (Fiona), Lee Sullivan (Charlie). Principal songs: "Waitin' For My Dearie,"

"I'll Go Home With Bonnie Jean," "The Heather On The Hill," "Come To Me, Bend To Me," "Almost Like Being In Love." 581 performances.

Allegro. Music by Richard Rodgers; book and lyrics by Oscar Hammerstein II. Majestic Theatre. October 10, 1947. John Battles (Joseph Taylor, Jr.), Annamary Dickey (Marjorie Taylor), William Ching (Dr. Taylor), Muriel O'Malley (Grandma Taylor), Lisa Kirk (Emily), Roberta Jonay (Jennie). Principal songs: "A Fellow Needs A Girl," "So Far," "The Gentleman Is A Dope." 315 performances.

My Romance. Music by Sigmund Romberg; book and lyrics by Rowland Leigh. Shubert Theatre. October 19, 1948. Lawrence Brooks (Rev. Armstrong), Anne Jeffreys (Mme. Cavallini). Principal songs: "Souvenirs," "Written In Your Hand." 95 performances.

South Pacific. Music by Richard Rodgers; book by Oscar Hammerstein II and Joshua Logan; lyrics by Hammerstein. Majestic Theatre. April 7, 1949. Mary Martin (Nellie), Ezio Pinza (Emile), William Tabbert (Lt. Cable), Juanita Hall (Bloody Mary), Betta St. John (Liat). Principal songs: "Some Enchanted Evening," "Bali Ha'i," "I'm Gonna Wash That Man Right Outa My Hair," "A Wonderful Guy," "Younger Than Springtime." 1,925 performances.

The King and I. Music by Richard Rodgers; book and lyrics by Oscar Hammerstein II. St. James Theatre. March 29, 1951. Gertrude Lawrence (Anna), Yul Brynner (The King), Dorothy Sarnoff (Lady Thiang), Doretta Morrow (Tuptim), Larry Douglas (Lun Tha). Principal songs: "I Whistle A Happy Tune," "Hello, Young Lovers," "Getting To Know You," "We Kiss In A Shadow," "Something Wonderful," "I Have Dreamed." 1,246 performances.

A Tree Grows in Brooklyn. Music by Arthur Schwartz; book by Betty Smith and George Abbott; lyrics by Dorothy Fields. Alvin Theatre. April 19, 1951. Marcia Van Dyke (Katie), Johnny Johnston (Johnny), Shirley Booth (Cissy). Principal songs: "Make The Man Love Me," "Look Who's Dancing," "He Had Refinement." 270 performances.

Paint Your Wagon. Music by Frederick Loewe; book and lyrics by Alan Jay Lerner. Shubert Theatre. November 12, 1951. James Barton (Ben), Olga San Juan (Jennifer), Tony Bavaar (Julio). Principal songs: "I Talk To The Trees," "I Still See Elisa," "They Call The Wind Maria," "Wandrin' Star." 289 performances.

Kismet. Music and lyrics by Robert Wright and George Forrest (music adapted from Borodin melodies); book by Charles Lederer and Luther Davis. Ziegfeld Theatre. December 3, 1953. Alfred Drake (Hajj), Doretta Morrow (Marsinah), Joan Diener (Lalume), Richard Kiley (Calif). Principal songs: "Stranger In Paradise," "Baubles, Bangles And Beads." 583 performances.

The Golden Apple. Music by Jerome Moross; book and lyrics by John Latouche. Phoenix Theatre. March 11, 1954. Priscilla Gillette (Penelope), Stephen Douglass (Ulysses), Kaye Ballard (Helen), Portia Nelson (Minerva).

Principal songs: "Lazy Afternoon," "It's The Going Home Together." 125 performances.

Fanny. Music and lyrics by Harold Rome; book by S. N. Behrman and Joshua Logan. Majestic Theatre. November 4, 1954. Ezio Pinza (César), Walter Slezak (Panisse), Florence Henderson (Fanny), William Tabbert (Marius). Principal songs: "Fanny," "Love Is A Very Light Thing." 888 performances.

My Fair Lady. Music by Frederick Loewe; book and lyrics by Alan Jay Lerner. Mark Hellinger Theatre. March 15, 1956. Rex Harrison (Henry Higgins), Julie Andrews (Eliza Doolittle), Michael King (Freddy), Stanley Holloway (Doolittle). Principal songs: "On The Street Where You Live," "The Rain In Spain," "I've Grown Accustomed To Her Face," "I Could Have Danced All Night," "Get Me To The Church On Time." 2,717 performances.

The Most Happy Fella. Music, book, and lyrics by Frank Loesser. Imperial Theatre. May 3, 1956. Robert Weede (Tony), Joe Sullivan (Rosabella), Art Lund (Joey), Susan Johnson (Cleo), Shorty Long (Herman). Principal songs: "Somebody, Somewhere," "The Most Happy Fella," "Standing On The Corner," "Big D." 676 performances.

Candide. Music by Leonard Bernstein; book by Lillian Hellman; lyrics by Richard Wilbur, Dorothy Parker, and John Latouche. Martin Beck Theatre. December 1, 1956. Robert Rounseville (Candide), Max Adrian (Dr. Pangloss), Barbara Cook (Cunegonde), Irra Petina (Old Lady). Principal songs: "The Best Of All Possible Worlds," "Glitter And Be Gay," "Make Our Garden Grow." [The overture is the most famous musical number.] 73 performances.

West Side Story. Music by Leonard Bernstein; book by Arthur Laurents; lyrics by Stephen Sondheim. Winter Garden Theatre. September 26, 1957. Carol Lawrence (Maria), Larry Kert (Tony), Chita Rivera (Anita). Principal songs: "Tonight," "Maria," "I Feel Pretty." 734 performances.

The Sound of Music. Music by Richard Rodgers; book by Howard Lindsay and Russel Crouse; lyrics by Oscar Hammerstein II. Lunt-Fontanne Theatre. November 16, 1959. Mary Martin (Maria), Theodore Bickel (Capt. Von Trapp), Patricia Neway (Mother Abbess). Principal songs: "The Sound Of Music," "My Favorite Things," "Climb Ev'ry Mountain," "Edelweiss." 1,443 performances.

Camelot. Music by Frederick Loewe; book and lyrics by Alan Jay Lerner. Majestic Theatre. December 3, 1960. Julie Andrews (Guenevere), Richard Burton (King Arthur), Robert Goulet (Lancelot). Principal songs: "Camelot," "If Ever I Would Leave You." 873 performances.

She Loves Me. Music by Jerry Bock; book by Joe Masteroff; lyrics by Sheldon Harnick. Eugene O'Neill Theatre. April 23, 1963. Barbara Cook (Amalia),

Daniel Massey (Georg), Jack Cassidy (Steven), Barbara Baxley (Ilona). Principal songs: "She Loves Me," "Dear Friend," "Ilona," "Ice Cream." 302 performances.

Fiddler on the Roof. Music by Jerry Bock; book by Joseph Stein; lyrics by Sheldon Harnick. Imperial Theatre. September 22, 1964. Zero Mostel (Tevye), Maria Karnilova (Golde), Beatrice Arthur (Yente). Principal songs: "Matchmaker, Matchmaker," "If I Were A Rich Man," "To Life," "Sunrise, Sunset." 3,242 performances.

Man of La Mancha. Music by Mitch Leigh; book by Dale Wasserman; lyrics by Joe Darion. ANTA Washington Sq. Theatre. November 22, 1965. Richard Kiley (Don Quixote), Joan Diener (Aldonza), Irving Jacobson (Sancho Panza), Ray Middleton (Innkeeper), Robert Rounseville (Padre). Principal songs: "Man Of La Mancha," "Dulcinea," "The Impossible Dream." 2,328 performances.

A Little Night Music. Music and lyrics by Stephen Sondheim; book by Hugh Wheeler. Shubert Theatre. February 25, 1973. Glynis Johns (Desirée), Len Cariou (Fredrik), Hermione Gingold (Mme. Armfeldt). Principal songs: "Send In The Clowns," "Liaisons." 601 performances.

Sweeney Todd. Music and lyrics by Stephen Sondheim; book by Hugh Wheeler. Uris Theatre. March 1, 1979. Len Cariou (Todd), Angela Lansbury (Mrs. Lovett), Victor Garber (Anthony), Ken Jennings (Tobias). Principal songs: "Johanna," "A Little Priest," "Not While I'm Around," "The Ballad Of Sweeney Todd." 557 performances.

Index